BASIC
Engineering
PRACTICES

Bruce J. Black C.ENG MIEE

Edward Arnold
A member of the Hodder Headline Group
LONDON   MELBOURNE   AUCKLAND

Acknowledgements

Chubb Fire Ltd	Figs 1.12, 1.13, 1.14, 1.15
Lloyds British Testing Ltd for information on lifting equipment	
Neill Tools Ltd	Figs 3.2, 4.17, 4.38, 4.46, 4.49, 5.9, 5.10, 5.13
Rubert & Co	Fig. 4.4
Mitutoyo (UK) Ltd	Figs 4.9, 4.12, 4.13, 4.14, 4.15, 4.16, 4.20, 4.21, 4.22, 4.23, 4.24, 4.26, 4.27, 4.29, 4.30, 4.31, 4.50, 5.14, 5.19
Thomas Mercer Ltd	Fig 4.28, 4.32
Pratt Burnerd International Ltd	Fig 6.12

First published in Great Britain 1995 by
Edward Arnold, a division of Hodder Headline PLC,
338 Euston Road, London NW1 3BH

British Library Cataloguing in Publication Data
A catalogue record for this book is available from the British Library

ISBN 0 340 60153 1

1 2 3 4 5   95 96 97 98 99

Produced by Gray Publishing, Tunbridge Wells
Printed and bound in Great Britain by The Bath Press, Avon

Contents

1

Safe practices

Almost everyone working in a factory has at some stage in his or her career suffered an injury requiring some kind of treatment or first aid. It may have been a cut finger or something more serious. The cause may have been carelessness by the victim or a colleague, defective safety equipment, not using the safety equipment provided, or inadequate protective clothing. Whatever the explanation given for the accident, the true cause was most likely a failure to think ahead. You must learn to work safely. Your workplace will have its own safety rules so obey them at all times. Ask if you don't understand any instruction and do report anything which seems dangerous, damaged or faulty.

Health and Safety at Work Act 1974 (HSW Act)

This Act of Parliament came into force in April 1975 and covers all people at work except domestic servants in a private household. It is aimed at people and their activities, rather than at factories and the processes carried out within them.

The purpose of the Act is to provide a legal framework to encourage high standards of health and safety at work.

Its aims are:

a) to secure the health, safety, and welfare of people at work;

b) to protect other people against risks to health or safety arising from the activity of people at work;

c) controlling the keeping and use of dangerous substances and preventing people from unlawfully having or using them;

d) controlling the release into the atmosphere of noxious or offensive substances, from prescribed premises.

Health and safety organisation (Fig. 1.1)

The HSW Act established two bodies, the Health and Safety Commission and the Health and Safety Executive.

Most of the health and safety regulations are the responsibility of the Secretary of State for Employment. These regulations are normally based on proposals submitted by the Health and Safety Commission after consultation with organisations representing, among others, employees, employers, local authorities, and professional bodies.

The Health and Safety Commission consists of representatives from both sides of industry, and from local authorities, and is responsible for developing policies in health and safety.

The Health and Safety Executive is appointed by the Commission with the approval of the Secretary of State and is responsible for enforcing legal requirements, as well as providing an advisory service to both sides of industry.

The Executive also appoints inspectors to carry out its enforcement functions.

Inspectors may visit a workplace without notice. They may want to investigate an

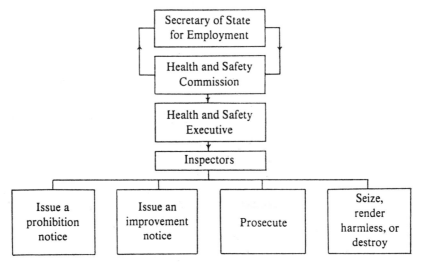

Figure 1.1 Health and safety organisation.

accident or complaint, or examine the safety, health and welfare aspects of the business. They have the right to talk to employees and safety representatives and to take photographs and samples.

If there is a problem an inspector can

a) issue a prohibition notice to stop any activity which could result in serious personal injury, until remedial action is taken;

b) issue an improvement notice requiring a fault to be remedied within a specified time;

c) prosecute any person who does not comply with the regulations – this can lead to a fine, imprisonment, or both;

d) seize, render harmless or destroy any substance or article considered to be the cause of imminent danger or serious personal injury.

Employers' responsibilities
(Fig. 1.2)

Employers have a general duty under the HSW Act 'to ensure, so far as is reasonably practicable, the health, safety and welfare at work of their employees'. The HSW Act specifies five areas which in particular are covered by the employers general duty:

1 To provide and maintain machinery, equipment and other plant, and systems of work that are safe and without risk to health.

('Systems of work' means the way in which the work is organised and includes, layout of the workplace, the order in which jobs are carried out, or special precautions to be taken before carrying out certain hazardous tasks.)

2 Ensure ways in which particular articles and substances (e.g. machinery and chemicals) are used, handled, stored and transported are safe and without risk to health.

3 Provide information, instruction, training and supervision necessary to ensure health and safety at work. **Information** means the background knowledge needed to put the instruction and training into context. **Instruction** is when someone shows others how to do something by practical demonstration. **Training** means having employees practise a task to improve their performance. **Supervision** is needed to oversee and guide in all matters related to the task.

4 Ensure any place under their control and where their employees work, is kept in a safe condition and does not pose a risk to health. This includes ways into and out of the workplace.

5 Ensure the health and safety of their employees working environment (e.g heating, lighting, ventilation, etc.) Must also provide adequate arrangements for welfare at work of their employees (the term 'welfare at work' covers facilities such as seating, washing, toilets, etc.)

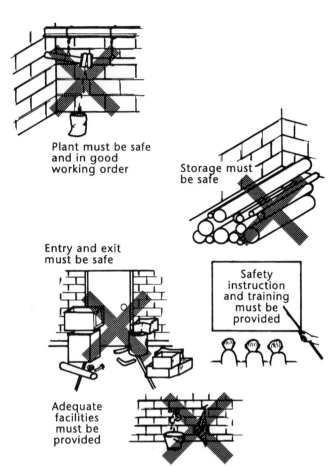

Plant must be safe and in good working order

Storage must be safe

Entry and exit must be safe

Safety instruction and training must be provided

Adequate facilities must be provided

Figure 1.2 Duties of employers.

Safety policy

The HSW Act requires every employer employing more than five people to prepare a written statement of their safety policy. The written policy statement must set out the employers' aims and objectives for improving health and safety at work.

The purpose of a safety policy is to ensure that employers think carefully about hazards at the workplace and about what should be done to reduce those hazards to make the workplace safe and healthy for their employees.

Another purpose is to make employees aware of what policies and arrangements are being made for their safety. For this reason you must be given a copy which you must read, understand and follow.

The written policy statement needs to be reviewed and revised jointly by employer and employees representatives as appropriate working conditions change or new hazards arise.

Safety Representatives and Safety Committees Regulations 1977

Safety representatives

The Regulations came into force on 1 October 1978 and provide recognised trade unions with the right to appoint safety representatives to represent the employees in consultations with their employers about health and safety matters of the organisation.

The HSW Act requires every employer to consult safety representatives in order to make and maintain arrangements to enable the employer and the employees to cooperate in the promotion and development of health and safety measures and to check their effectiveness.

An employer must give safety representatives the necessary time off, with pay, to carry out their functions and receive appropriate training.

The functions of a safety representative include:

a) to investigate potential hazards and dangerous occurrences in the workplace;
b) to investigate complaints relating to an employees' health, safety or welfare at work
c) to make representations to the employer on matters affecting the health, safety or welfare of employees at the workplace;
d) to carry out inspections of the workplace where there has been a change in conditions of work, or there has been a notifiable accident or dangerous occurrence in a workplace or a notifiable disease has been contracted there;
e) to represent the employees he or she was appointed to represent in consultation with inspectors or any enforcing authority;
f) to attend meetings of safety committees.

Safety committees

The HSW Act requires an employer to establish a safety committee if requested in writing by at least two safety representatives. The main objective of such a committee is to promote cooperation between employers and employees in setting up, developing and carrying out measures to ensure the health and safety at work of the employees. Its functions can include:

a) studying safety and accident reports so that unsafe and unhealthy conditions and practices may be identified and recommendations made for corrective action;
b) consider reports provided by inspectors and by safety representatives;
c) assist in developing works safety rules and safe systems of work;
d) monitor the effectiveness of employee safety training
e) monitor the adequacy of health and safety communication and publicity in the workplace;
f) provide a link with the appropriate enforcing agency.

Employees' responsibilities
(Fig. 1.3)

Under the HSW Act it is the duty of every employee while at work:

1 To take reasonable care for their own health and safety and that of others who may be affected by what they do or don't do.

   This duty implies not only avoiding silly or reckless behaviour but also understanding hazards and complying with safety rules and procedures. This means that you correctly use all work items provided by your employer in accordance with the training and instruction you received to enable you to use them safely.
2 To cooperate with their employer on health and safety.

   This duty means that you should inform, without delay, of any work situation which might be dangerous and notify any shortcomings in health and safety arrangements so that remedial action may be taken.

Every employee must take reasonable care of the safety of himself and others

Never interfere with anything provided in the interests of safety

Figure 1.3 Duties of employees.

The HSW Act also imposes a duty on all people, both people at work and members of the public, including children to not intentionally interfere with or misuse anything that has been provided in the interests of health, safety and welfare.

The type of things covered include fire escapes and fire extinguishers, perimeter fencing, warning notices, protective clothing, guards on machinery and special containers for dangerous substances.

You can see that it is essential for you to adopt a positive attitude and approach to health and safety in order to avoid, prevent and reduce risks at work. Your training is an important way of achieving this and contributes, not only your own, but to the whole organisation's health and safety culture.

New regulations for health and safety at work

Six new sets of health and safety at work regulations came into force on 1 January 1993. The new regulations implement European Community (EC) directives on health and safety at work in the move towards a single European Union. At the same time they are part of a continuing modernisation of existing UK law.

Most of the duties in the new regulations are not completely new but clarify and make more explicit what is in current health and safety law. A lot of out-of-date law will be repealed by the new regulations, for example many parts of the Factories Act 1961.

The six regulations are:

- Management of Health and Safety at Work Regulations 1992;
- Provision and Use of Work Equipment Regulations 1992;
- Workplace (Health, Safety and Welfare) Regulations 1992;
- Personal Protective Equipment at Work Regulations 1992;
- Health and Safety (Display Screen Equipment) Regulations 1992 (covers computer monitors and is not relevant to this book);
- Manual Handling Operations Regulations 1992 (see page 19).

Management of Health and Safety at Work Regulations 1992

These Regulations set out broad general duties which operate with the more specific ones in other health and safety regulations. They are aimed mainly at improving health and safety management. Their main provisions are designed to encourage a more systematic and better organised approach to dealing with health and safety.

The Regulations require employers to:

a) assess the risk to health and safety of employees and anyone else who may be affected so that the necessary preventive and protective measures can be identified;
b) make arrangements for putting into practice the health and safety measures that follow from the risk assessment;
c) provide appropriate health surveillance of employees where necessary;
d) appoint competent people to help devise

and apply the measures needed;
e) set up emergency procedures;
f) give employees information about health and safety matters;
g) make sure that employees have adequate health and safety training and are capable enough at their jobs to avoid risk;
h) co-operate with any other employers who share a work site;
i) give some health and safety information to temporary workers, to meet their special needs.

The Regulations also:

a) places a duty on employees to follow health and safety instructions and report danger; and
b) extends the current law which requires employers to consult employees safety representatives and provide facilities for them.

Provision and Use of Work Equipment Regulations 1992 (PUWER)

These Regulations lay down important health and safety laws for the provision and use of work equipment and is designed to pull together and tidy up the laws governing equipment used at work. Much old legislation including seven sections of the Factories Act 1961 will be replaced. Its primary objective is to ensure the provision of safe work equipment and its safe use.

These Regulations came into force on 1 January 1993 and will operate alongside the HSW Act. Some of the Regulations do not apply to certain categories of work equipment until 1 January 1997.

Work equipment has wide meaning and is broadly defined to include anything from a hand tool, through machines of all kinds, to a complete plant such as a refinery.

PUWER cover the health and safety requirement in respect of the following:

1 The suitability of work equipment – equipment must be suitable by design and construction for the actual work it is provided to do.

2 Maintenance of work equipment in good repair – from simple checks on hand tools such as loose hammer heads to specific checks on lifts and hoists. When maintenance work is carried out it should be done in safety and without risk to health.

3 Information and instruction on use of the work equipment including instruction sheets, manuals or warning labels from manufacturers or suppliers. Adequate training for the purposes of health and safety in the use of specific work equipment.

4 Dangerous parts of machinery – guarding machinery to avoid the risks arising from mechanical hazards. The principal duty is to take effective measures to prevent contact with dangerous parts of machinery by providing:
   ■ fixed enclosing guards;
   ■ other guards (see Fig. 1.4) or protection devices;
   ■ protection appliances (jigs, holders);
   ■ information, instruction, training and supervision.

5 Protection against specified hazards:
   ■ material falling from equipment;
   ■ material ejected from a machine;
   ■ parts of the equipment breaking off e.g. grinding wheel bursting;
   ■ parts of equipment collapsing e.g. scaffolding;
   ■ overheating or fire e.g. bearing running hot, ignition by welding torch;
   ■ explosion of equipment e.g. failure of a pressure-relief device;
   ■ explosion of substance in the equipment e.g. ignition of dust.

6 High and very low temperature – prevent the risk of injury from contact with hot (blast furnace, steam pipes) or very cold work equipment (cold store).

7 Controls and control systems – starting work equipment should only be possible by using a control and it should not be possible to be accidentally or inadvertently operated nor 'operate itself' (by vibration or failure of a spring mechanism).

Stop controls should bring the equipment to a safe condition in a safe manner. Emergency stop controls are intended to effect a rapid response to potentially

Figure 1.4 Guard fitted to horizontal milling machine.

dangerous situations and should be easily reached and actuated. Common types are mushroom headed buttons (see Fig. 1.5), bars, levers, kick plates or pressure-sensitive cables.

It should be possible to identify easily what each control does. Both the controls and their markings should be clearly visible and factors such as colour, shape and position are important.

8 Isolation from source of energy – to allow equipment to be made safe under particular circumstances, for example when maintenance is to be carried out or when an unsafe condition develops. Isolation may be achieved by simply removing a plug from an electrical socket or by operating an isolating switch or valve.

Figure 1.5 Mushroom-headed stop button.

Sources of energy may be electrical, pressure (hydraulic or pneumatic) or heat.

9 Stability – there are many types of work equipment that might fall over, collapse or overturn unless they are fixed. Most machines used in a fixed position should be bolted down. Some types of work equipment such as mobile cranes may need counterbalance weights.

Ladders should be at the correct angle (a slope of four units up to each one out from the base), correct height (at least 1 metre above the landing place) and tied at the top or secured at the foot.

10 Lighting – if the lighting in the workplace is insufficient for detailed tasks then additional lighting will need to be provided, for example. local lighting on a machine (Fig. 1.6).

11 Markings – there are many instances where marking of equipment is appropriate for health and safety reasons, for example start/stop controls, safe working load on cranes or types of fire extinguishers (see page 15).

12 Warnings – normally in the form of a permanent printed notice or similar, for example: 'head protection must be worn' (see page 41). Portable warnings are also necessary during temporary operations such as maintenance.

Warning devices can be used which may be audible, for example reversing alarms on heavy vehicles, or visible, for example lights on a control panel. They may indicate imminent danger, development of a fault or the continued presence of a potential hazard.

They must all be easy to see, understand and be unambiguous.

Workplace (Health, Safety and Welfare) Regulations 1992

These Regulations will also tidy up a lot of existing requirements. They will replace many pieces of old law, including parts of the Factories Act 1961. They will be much easier to understand making it clearer what is expected of everyone. They came into force on 1 January 1993 but for existing workplaces the Regulations take effect on 1 January 1996.

These Regulations set general requirements which are listed here in four broad areas:

■ **Working environment**
  • ventilation
  • temperature in indoor workplace
  • lighting including emergency lighting
  • room dimensions and space
  • suitability of workstations and seating.

■ **Safety**
  • safe passage of pedestrians and vehicles (e.g. traffic routes, must be wide enough and marked where necessary, and there must be enough of them)
  • windows and skylights (safe opening, closing and cleaning)
  • transparent or translucent surfaces in doors and partitions (use of safety material and marking)
  • doors, gates and escalators (safety devices)
  • floors (construction and maintenance, obstructions and slipping and tripping hazards)
  • falling from heights and into dangerous substances
  • falling objects.

■ **Facilities**
  • toilets
  • washing, eating and changing facilities

Figure 1.6 Local lighting on a centre lathe.

- clothing storage
- drinking water
- rest areas (and arrangements to protect people from the discomfort of tobacco smoke).

■ **Housekeeping**
  - maintenance of workplace, equipment and facilities
  - cleanliness
  - removal of waste materials.

Personal Protective Equipment at Work Regulations 1992

These Regulations came into force on 1 January 1993 and set out in legislation, sound principles of selecting, providing and using personal protective equipment (PPE). They replace parts of over 20 old pieces of law (e.g. the Protection of Eyes Regulations 1974 has been revoked). They do not replace the recently introduced laws dealing with PPE (e.g. Control of Substances Hazardous to Health or Noise at Work Regulations).

PPE should always be relied upon as a last resort to protect against risks to health and safety. Engineering controls and safe systems of work should always be considered first. Where the risks are not adequately controlled by other means, the employer has a duty to ensure that suitable PPE is provided, free of charge. PPE will only be suitable if it is appropriate for the risks and the working conditions; takes account of the workers needs and fits properly; gives adequate protection; and is compatible with any other item of PPE worn.

The employer also has duties to:

a) assess the risks and PPE intended to be issued and that it is suitable;
b) maintain, clean and replace PPE;
c) provide storage for PPE when it is not being used;
d) ensure that PPE is properly used; and
e) give training, information and instruction to employees on the use of PPE and how to look after it.

PPE is defined as all equipment which is intended to be worn or held to protect against

risk to health and safety. This includes most types of protective clothing and equipment such as: eye, head, foot and hand protection; and protective clothing for the body. It does not include ear protectors and respirators which are covered by separate existing regulations.

Eye protection: serves as a guard against the hazards of impact, splashes from chemicals or molten metal, liquid droplets (chemical mists and sprays), dust, gases and welding arcs. Eye protectors include safety spectacles, eye-shields, goggles, welding filters, face shields and hoods (Fig. 1.7).

Head protection: includes industrial safety helmets to protect against falling objects or impact with fixed objects; industrial scalp protectors to protect against striking fixed

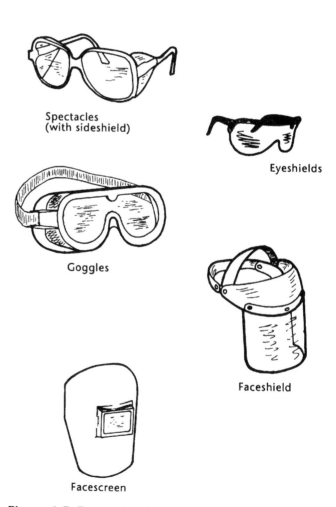

Spectacles
(with sideshield)

Eyeshields

Goggles

Faceshield

Facescreen

Figure 1.7 Eye protection.

obstacles, scalping or entanglement; and caps and hairnets to protect against scalping and entanglement.

Foot protection: includes safety boot or shoe with steel toe caps; foundry boots with steel toe caps, which are heat resistant and designed to keep out molten metal; wellington boots to protect against water and wet conditions; and anti-static footwear to prevent the build up of static electricity on the wearer.

Hand protection: gloves of various design provide protection against a range of hazards including cuts and abrasions; extremes of temperature (hot and cold); skin irritation and dermatitis; and contact with toxic or corrosive liquids. Barrier creams may sometimes be used as an aid to skin hygiene in situations where gloves cannot be used.

Protective clothing: types of clothing used for body protection include coveralls, overalls and aprons to protect against chemicals and other hazardous substances; outfits to protect against cold, heat and bad weather; and clothing to protect against machinery such as chain saws. Types of clothing worn on the body to protect the person include high visibility clothing; life-jackets and buoyancy aids.

The Reporting of Injuries, Diseases and Dangerous Occurrences Regulations 1985 (RIDDOR)

RIDDOR came into effect in April 1986. These Regulations require injuries, diseases and occurrences in specified categories to be notified to the relevant enforcing authority. In the case of a factory, the enforcing authority is the Health and Safety Executive.

The enforcing authority must be notified without delay, normally by a phone call, followed by a written report within seven days.

Immediate notification is required for the following:

a) any fatal injuries to employees or other people in an accident connected with your business;
b) any major injuries to employees or other people in an accident connected with your business (major injuries include fractures, amputation, loss of sight, injury from electric shock and any other injury which results in the person being admitted to hospital for more than 24 hours). A written report must be sent within seven days of any other injury to an employee which results in an absence of more than three working days;
c) any of the dangerous occurrences listed in the Regulations (these include the collapse, overturning or failure of lifts, hoists and cranes, explosion of vessels, electrical fires, the sudden release of highly flammable liquids);
d) report notifying specific diseases related to particular work activities listed in the Regulations (the general diseases covered include certain poisonings, some skin diseases, lung diseases, infections and other conditions such as occupational cancer).

A record must be kept of any injury, occurrence or case of disease requiring report. This should include the date, time and place, personal details of those involved and a brief description of the nature of the event.

Noise at Work Regulations 1989

These Regulations are intended to reduce hearing damage caused by loud noise. Exposure to high noise levels can cause incurable hearing damage. The important factors are: (1) the noise level, given in decibel units dB(A) and (2) how long the person is exposed to the noise, daily, or over a number of years.

Action levels are set and action has to be taken when they are reached:

- First action level 85 dB(A)
- Second action level 90 dB(A). These are personal daily exposure levels and are denoted $L_{EP,d}$.
- Peak action level equivalent to 140 dB(A) (where cartridge-operated tools are used even occasionally). Example levels are shown in Table 1.1.

The need to wear ear protection should be the last resort. The best protection against

Table 1.1

Noise source	Noise level dB(A)
Domestic food blender	81
Electric drill	87
Sheet metal shop	93
Circular saw	99
Chain saw	102
Hand grinding metal	108
Jet aircraft taking off	
25 metres away	140

noise is to control it at source by designing or choosing machines and processes to make less noise, by enclosing noisy machines or putting in a separate room, or by fitting silencers. Finally, if all else fails, ear protection should be provided in noisy areas.

The Regulations require an employer to:

■ assess noise levels;
■ inform workers at the first action level about the risks to hearing and provide ear protectors;
■ control noise exposure if noise reaches second or peak action levels;
■ mark ear protection zones with notices and make sure that everyone wears ear protectors.

Electrical hazards

Electrical equipment of some kind is used in every factory. Electricity should be treated with respect – it cannot be seen or heard, but it can kill. Even if it is not fatal, serious disablement can result through shock and burns. Also, a great deal of damage to property and goods can be caused, usually through fire or explosion as a result of faulty wiring or faulty equipment.

The Electricity at Work Regulations 1989 came into force on 1 April 1990. The purpose of the Regulations is to require precautions to be taken against the risk of death or personal injury from electricity in work activities.

The Institution of Electrical Engineers Regulations for electrical installations (IEE Wiring Regulations), although non-statutory, is widely recognised and accepted in the UK and compliance with these is likely to comply with the relevant parts of the Electricity at Work Regulations 1989.

The major hazards arising from the use of electrical equipment are:

Electric shock: the body responds in a number of ways to electric current flowing through it, any one of which can be fatal. The chance of electric shock is increased in wet or damp conditions, or close to conductors such as working in a metal tank. Hot environments where sweat or humidity reduce the insulation protection offered by clothing increases the risk.

Electric burn: this is due to the heating effect caused by electric current passing through body tissue, most often the skin at the point of contact giving rise to the electric shock.

Fire: caused by electricity in a number of ways including; overheating of cables and electrical equipment due to overloading; leakage currents due to poor or inadequate insulation; overheating of flammable materials placed too close to electrical equipment; ignition of flammable materials by sparking of electrical equipment.

Arcing: generates ultra-violet radiation causing a particular type of burn similar to severe sunburn. Molten metal resulting from arcing can penetrate, burn and lodge in the flesh. Ultra-violet radiation can also cause damage to sensitive skin and to eyes, e.g. arc eye in metal arc welding.

Explosion: these include the explosion of electrical equipment, e.g. switchgear or motors, or where electricity causes the ignition of flammable vapours, gases, liquids and dust by electric sparks or high temperature electrical equipment.

Electrical precautions

Where it is possible for electrical equipment to become dangerous if a fault should arise, then precautions must be taken to prevent injury. These precautions include:

Double insulation: the principle is that the live conductors are covered by two discrete layers of insulation. Each layer would provide adequate insulation in themselves but together they ensure little likelihood of danger arising from insulation failure. This arrangement avoids the need for an earth wire. Double insulation is particularly suitable for portable equipment such as drills. However, safety depends on the insulation remaining in sound condition and the equipment must be properly constructed, used and maintained.

Earthing: in the UK the electricity supply is connected to earth. It is this system that enables earth faults on electrical equipment to be detected and the electrical supply to be cut off automatically. This automatic cut-off is performed by fuses or automatic circuit breakers: if a fault occurs the fuse will blow and break the circuit. Although they do not eliminate the risk of electric shock, danger may be reduced by the use of a residual current device (RCD) designed to operate rapidly at small leakage currents. RCDs should only be considered as providing a second line of defence. It is essential to regularly operate the test trip button to maintain their effectiveness.

Use of safe voltages: reduced voltage systems (110 volts) are particularly suitable for portable electrical equipment in construction work and in high conducting locations such as boilers, tunnels, tanks; where the risk to equipment and trailing cables is high; and where the body may be damp.

The human body as part of a circuit

In order to minimise the risk of shock and fire, any metalwork other than the current-carrying conductor must be connected to earth. The neutral of the electrical supply is earthed at the source of distribution, i.e. the supply transformer, so that, if all appliances are also connected to earth, a return path for the current will be available through earth when a fault occurs (see Fig. 1.8). To be effective, this earth path must be of sufficiently low resistance to pass a relatively high current when a fault occurs. This higher current will in turn operate the safety device in the circuit, i.e. the fuse will blow.

Accidents happen when the body provides a

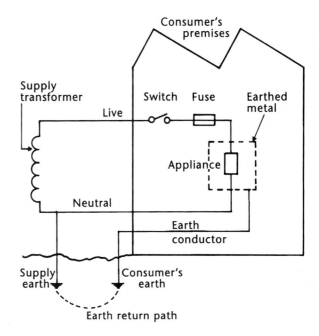

Figure 1.8 Electric circuit for premises.

direct connection between the live conductors – when the body or a tool touches equipment connected to the supply. More often, however, the connection of the human body is between one live conductor and earth, through the floor or adjacent metalwork (see Fig. 1.9). Metal pipes carrying water, gas, or steam, concrete floors, radiators, and machine structures all readily provide a conducting path of this kind.

Any article of clothing containing any metal parts increases the likelihood of accidental electrical contact. Metal fittings such as buttons, buckles, metal watch or wrist

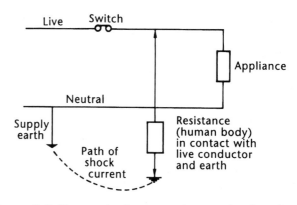

Figure 1.9 Human body as a resistance in electric circuit.

bracelets or dog tags, and even rings could result in shock or burns.

Wetness or moisture at surfaces increases the possibility of leakage of electricity, by lowering the resistance and thus increasing the current. Contact under these conditions therefore increases the risk of electric shock.

All metals are good electrical conductors and therefore all metallic tools are conductors. Any tool brought near a current carrying conductor will bring about the possibility of a shock. Even tools with insulated handles do not guarantee that the user will not suffer shock or burns.

Electric shock and treatment

If the human body accidentally comes in contact with an electrical conductor which is connected to the supply, a current may, depending on the conditions, flow through the body. This current will at least produce violent muscular spasms which may cause the body to be flung across the room or fall off a ladder. In extreme cases the heart will stop beating.

Burns are caused by the current acting on the body tissue and internal heating can aiso take place leading to partial blockage of blood flow through the blood vessels.

In the event of someone suffering electric shock, know what to do – it should form part of your training.

1 Shout for help – if the casualty is still in contact with electric current, switch off or remove the plug.
2 If the current cannot be switched of, take special care to stand on a dry non-conducting surface and pull or push the victim clear using a length of dry cloth, jacket, or a broom. Remember: do not touch the casualty as you will complete the circuit and also receive a shock.
3 Once free, if the casualty is breathing, put in recovery position and get the casualty to hospital; if the casualty is not breathing give mouth-to-mouth resuscitation, check pulse, and, if absent, apply chest compressions and call for medical assistance.

Posters giving detailed procedure to be followed in the event of a person suffering electric shock must be permanently displayed in your workplace. With this and your training you should be fully conversant with the procedures – remember it could save a life.

General electrical safety rules

■ Ensure that a properly wired plug is used for all portable electrical equipment (see Fig. 1.10)
  • **brown** wire   **live** conductor
  • **blue** wire   **neutral** conductor
  • **green/yellow** wire **earth** conductor.
■ Never improvise by jamming wires in sockets with nails or matches.
■ Moulded rubber plugs are preferable to the brittle plastic types, since they are less prone to damage.
■ All electrical connections must be secure, loose wires or connections can arc.
■ A fuse of the correct rating must be fitted – this is your safeguard if a fault develops – never use makeshift fuses such as pieces of wire.
■ Any external metal parts must be earthed so that if a fault develops, the fuse will blow and interrupt the supply.
■ Never run power tools from lamp sockets.
■ Connection between the plug and equipment should be made with the correct cable suited to the current rating of the equipment.
■ Old or damaged cable should never be used.
■ Equipment should always be disconnected from the mains supply before making any adjustment, even when changing a lamp.
■ Do not, under any circumstances, interfere with any electrical equipment or attempt to repair it yourself. All electrical work should

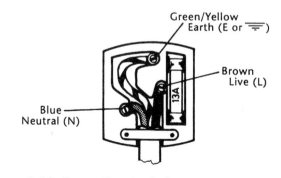

Figure 1.10 Correctly wired plug.

be done by a qualified electrician. A little knowledge is often sufficient to make electrical equipment function but a much higher level of knowledge and expertise is usually needed to ensure safety.

Fire

Fire is a phenomenon in which combustible materials, especially organic materials containing carbon, react chemically with the oxygen in the air to produce heat. Flame arises from the combustion of volatile liquids and gases evolved and spreads the fire.

No one should underestimate the danger of fire. Many materials burn rapidly and the fumes and smoke produced, particularly from synthetic material, including plastics, may be deadly.

There are a number of reasons for fires starting:

■ malicious ignition: i.e. deliberate fire raising;
■ misuse or faulty electrical equipment: e.g. incorrect plugs and wiring, damaged cables, overloaded sockets and cables, sparking, and equipment such as soldering irons left on and unattended;
■ cigarettes and matches: smoking in unauthorised areas, throwing away lighted cigarettes or matches;
■ mechanical heat and sparks: e.g. faulty motors, overheated bearings, sparks produced by grinding and cutting operations;
■ heating plant: flammable liquids/substances in contact with hot surfaces;
■ rubbish burning: casual burning of waste and rubbish.

There are a number of reasons for the spread of fire including:

■ delayed discovery;
■ presence of large quantities of combustible materials;
■ lack of fire separating walls between production and storage areas;
■ openings in floors and walls between departments;
■ rapid burning of dust deposits;
■ oils and fats flowing when burning;
■ combustible construction of buildings;
■ combustible linings of roofs, ceilings and walls.

Fire prevention

The best prevention is to stop a fire starting:

■ where possible use materials which are less flammable;
■ minimise the quantities of flammable materials kept in the workplace or store;
■ store flammable material safely, well away from hazardous processes or materials, and where appropriate, from buildings;
■ warn people of the fire risk by a conspicuous sign at each workplace, storage area and on each container;
■ some items, like oil-soaked rags, may ignite spontaneously. Keep them in a metal container away from other flammable material;
■ before welding or similar work remove or insulate flammable material and have fire extinguishers to hand;
■ control ignition sources, e.g. naked flames and sparks, and make sure that 'no smoking' rules are obeyed;
■ do not leave goods or waste to obstruct gangways, exits, stairs, escape routes and fire points;
■ make sure that vandals do not have access to flammable waste materials;
■ comply with the specific precautions for highly flammable liquids e.g. petrol and flammable gas cylinders such as acetylene;
■ after each spell of work, check the area for smouldering matter or fire;
■ burn rubbish in a suitable container well away from buildings and have fire extinguishers to hand;
■ never wedge open fire-resistant doors designed to stop the spread of fire and smoke;
■ have enough fire extinguishers, of the right type and properly maintained, to deal promptly with small outbreaks of fire.

Fire Precautions Act 1971

Under this Act the owner of a factory employing more than 20 people (more than ten if on a floor above ground level) is required to have a fire certificate from the fire authority which specifies:

■ the use of the premises;
■ the means of escape in case of fire;

■ the fire-fighting equipment;
■ the fire-warning arrangements;

and may also include requirements for:

■ maintaining the means of escape and ensuring that they are not obstructed;
■ ensuring that employees receive instructions and training in what to do in the case of fire and keeping records of this training;
■ limiting the number of persons who may be in the premises at any one time.

Training has a most important bearing of the safety of the occupants of premises in the event of a fire and it may also contribute to reducing the extent of the damage. Training in fire prevention may be responsible for preventing a fire from starting.

Every employee of a firm should be trained:

■ to prevent fires;
■ in the action to take if fire occurs.

To ensure that all employees, after training, are familiar with, and understand the procedure in the event of a fire, repeated practice is desirable. After initial practices to establish the procedure, practice drills should be held at least twice a year.

Two-and-a-half minutes to complete evacuation is a reasonable standard to aim for, but in factories where there is a danger of rapid fire spread or of explosion, evacuation may need to be completed in less than one minute.

To avoid panic and accidents, fire drills should always be announced.

Fire fighting

Every employee should know where the portable fire extinguishers, the hose reels and the controls for extinguishing are located and how to operate extinguishers in their working area. This training must include the use of extinguishers on simulated fires.

It must be stressed that fire fighting should only be attempted if it is safe to do so and that an escape route must always be available.

It is also essential to emphasise the limits of first-aid fire fighting in order to show the need to attempt this safely and the importance of first raising the alarm.

As previously stated, a fire requires fuel, oxygen (air) and heat. This is shown by the 'fire triangle' in Fig. 1.11, where one side stands for fuel, another for heat and the third for air or oxygen. If any one side is removed the fire inside will go out.

The extinguishing of a fire is generally brought about by depriving the burning substances of oxygen and by cooling them to a temperature below which the reaction is not sustained.

By far the most important extinguishing agent, by reason of its availability and general effectiveness, is water. It is more effective than any other common substance in absorbing heat, thereby reducing the temperature of the burning mass. The steam produced also has a smothering action by lowering the oxygen content of the atmosphere near the fire.

For these reasons the use of a water hose reel in factories is common and is suitable for most fires except those involving flammable liquids or live electrical equipment.

For all practical purposes there are three main classes of fires: A, B and C as well as fires involving electrical equipment and those involving vehicles.

Class A type fires: fires involving combustible materials such as wood, paper and fabrics.

Class B type fires: fires involving flammable liquids such as oils, spirits, alcohols, greases, fats and certain plastics.

Class C type fires: fires involving flammable gases such as propane and butane.

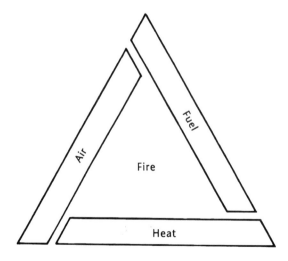

Figure 1.11 Fire triangle.

Types of portable fire extinguishers

Water (Fig. 1.12) – colour coded red – these are suitable for class A type fires. Water is a fast, efficient means of extinguishing these materials and works by having a rapid cooling effect on the fire so that insufficient heat remains to sustain burning and continuous ignition ceases.

Spray foam (Fig. 1.13) – colour coded ivory – these are ideal in multi-risk situations where both class A and B type fires are likely. Spray foam has a blanketing effect which both smothers the flame and prevents re-ignition of flammable vapours by sealing the surface of the material. These extinguishers contain an aqueous film-forming foam (AFFF).

Dry powder (Fig. 1.14) – colour coded blue – these are suitable for class A, B and C type fires and for vehicle protection. Because dry powder is non-conductive it is ideal for electrical hazards. Dry powder is a highly effective means of extinguishing fires as it interferes with the combustion process and provides rapid fire knockdown.

Figure 1.13 Spray foam fire extinguisher.

Carbon dioxide (CO_2 gas) (Fig. 1.15) – colour coded black – these are suitable for class B and C type fires. They are also ideal for electrical hazards because CO_2 is non-conductive. CO_2 is an extremely fast fire control medium. These extinguishers deliver a powerful concentration of carbon dioxide gas under great pressure, which smothers the flames very rapidly by displacing air from the local area of the fire.

Figure 1.12 Water fire extinguisher.

Figure 1.14 Dry powder fire extinguisher.

Figure 1.15 Carbon dioxide (CO$_2$ gas) fire extinguisher.

CO$_2$ is a non-toxic, non-corrosive gas that is harmless to most delicate equipment and materials found in situations such as computer rooms.

Halon – because of the potential ozone depleting effect, halon portable fire extinguishers are now being withdrawn.

Table 1.2 shows the portable extinguisher most suited to each class of fire.

Table 1.2

Class of fire	Type of extinguisher			
	Water	Spray foam	Dry powder	CO$_2$ gas
Type A paper, wood, fabric	✓	✓	✓	
Type B flammable liquids		✓	✓	✓
Type C flammable gases			✓	✓
Electrical hazards			✓	✓
Vehicle protection		✓	✓	

Causes of accidents

Workplace accidents can be prevented – you only need commitment, commonsense and follow the safety rules set out for your workplace. Safety doesn't just happen – you have to make it happen.

Most accidents are caused by carelessness, it may be you or a colleague, through failure to think ahead or as a result of fatigue. Fatigue may be brought on by working long hours without sufficient periods of rest or even through doing a second job in the evening.

Taking medicine can affect peoples ability to work safely, as can the effects of alcohol. Abuse of drugs or substances such as solvents can also cause accidents at work.

Serious injury and even death have resulted from horseplay, practical jokes or silly tricks. There is no place for this type of behaviour in the workplace.

Improper dress has led to serious injury: wearing trainers instead of safety footwear, and loose cuffs, torn overalls, floppy woollen jumpers, rings, chains, watch straps and long hair to get tangled up.

Don't forget, quite apart from the danger to your own health and safety, you are breaking the law if you fail to wear the appropriate personal protective equipment.

Unguarded or faulty machinery, and tools are other sources of accidents. Again within the health and safety law you must not use such equipment and furthermore it is your duty to immediately report defective equipment.

Accidents can occur as a result of the workplace environment, e.g. if ventilation is poor, temperature too high or too low, bad lighting, unsafe passages, doors, floors, and dangers from falls and falling objects.

Also if the workplace, equipment and facilities are not maintained, are not clean, and rubbish and waste materials are not removed.

Many accidents befall new workers in an organisation especially the young, and is the result of inexperience, lack of information, instruction, training or supervision all of which is the duty of the employer to provide.

General health and safety precautions

As already stated you must adopt a positive attitude and approach to health and safety. Your training is an important way of achieving competence and helps to convert information into healthy and safe working practices.

Remember to observe the following precautions:

Horseplay:
- work is not the place for horseplay, practical jokes, or silly tricks.

Hygiene:
- always wash your hands using suitable hand cleaners and warm water before meals, before and after going to the toilet, and at the end of each shift;
- dry you hands carefully on the clean towels or driers provided. Don't wipe them on old rags;
- paraffin, petrol or similar solvents should never be used for skin-cleaning purposes;
- use appropriate barrier cream to protect your skin;
- conditioning cream may be needed after washing to replace fatty matter and prevent dryness.

Housekeeping:
- never throw rubbish on the floor;
- keep gangways and work areas free of metal bars, components, etc;
- if oil, water, or grease is spilled, wipe it up immediately or someone might slip and fall.

Moving about:
- always walk – never run;
- keep to gangways – never take shortcuts;
- look out for and obey warning notices and safety signs;
- never ride on a vehicle not made to carry passengers, e.g. fork-lift trucks.

Personal protective equipment:
- use all personal protective clothing and equipment, such as ear and eye protectors, dust masks, overalls, gloves, safety shoes and safety helmets;
- get replacements if damaged or worn.

Ladders:
- do not use ladders with damaged, missing or loose rungs;
- always position ladders on a firm base and at the correct angle – the height of the top support should be four times the distance out from the base;
- ensure the ladder is long enough – at least one metre above the landing place;
- make sure the ladder is tied at the top or secured at the bottom;
- never over-reach from a ladder – be safe, get down and move it;
- take all necessary precautions to avoid vehicles or people hitting the bottom of the ladder.

Machinery:
- ensure you know how to stop a machine before you set it in motion;
- keep your concentration while the machine is in motion;
- never leave your machine unattended while it is in motion;
- take care not to distract other machine operators;
- never clean a machine while it is in motion – always isolate it from the power supply first;
- never clean swarf away with your bare hands – always use a suitable rake;
- keep your hair short or under a cap or hairnet – it can become tangled in drills or rotating shafts;
- avoid loose clothing – wear a snug-fitting boiler suit, done up, and ensure that any neckwear is tucked in and secure;
- do not wear rings, chains, or watches at work – they have caused serious injury when caught accidentally on projections;
- do not allow unguarded bar to protrude beyond the end of a machine, e.g. in a centre lathe;
- always ensure that all guards are correctly fitted and in position – remember, guards are fitted on machines to protect you and others from accidentally coming in contact with dangerous moving parts.

Harmful substances:
- learn to recognise hazard warning signs and labels;
- follow all instructions;
- before you use a substance find out what to do if it spills onto your hands or clothes;
- never eat or drink in the near vicinity;
- do not take home any clothes which have become soaked or stained with harmful substances;

■ do not put liquids or substances into unlabelled or wrongly labelled bottles or containers.

Electricity:

■ make sure you understand all instructions before using electrical equipment;

■ do not use electrical equipment for any other purpose nor in any area other than which it was intended;

■ always switch off or isolate before connecting or disconnecting any electrical equipment.

Compressed air:

■ only use compressed air if allowed to do so;

■ never use compressed air to clean a machine – it may blow in your face or someone else's and cause an injury.

Fire:

■ take care when using flammable substances;

■ never smoke in 'no smoking' areas;

■ do not throw rubbish, cigarette ends, and matches in corners or under benches;

■ always make sure that matches and cigarettes are put out before throwing them away;

■ know the correct fire drill.

First aid:

■ have first aid treatment for every injury however trivial;

■ know the first aid arrangements for your workplace.

2

Moving loads

More than a quarter of accidents reported each year are associated with manual handling. Manual handling is defined as the transporting or supporting of loads by hand or bodily force, i.e. human effort as opposed to mechanical handling such as a crane or fork-lift truck. Introducing mechanical assistance, e.g. a sack truck, may reduce but not eliminate manual handling since human effort is still required to move, steady or position the load.

The load may be moved or supported by the hands or any other part of the body, e.g. the shoulders.

A load is defined as a distinct moveable object.

Common accidents are sprains or strains often to the back which arise from incorrect application and/or prolonged use of body force. A full recovery is not always made and can result in physical impairment or even permanent disability.

It is now widely accepted that an ergonomic approach will remove or reduce the risk of manual handling injury. Ergonomics can be described as 'fitting the job to the person rather than the person to the job'. This ergonomic approach looks at manual handling taking account of a whole range of relevant factors including the nature of the task, the load, the working environment and the capability of the individual.

The Manual Handling Operations Regulations 1992

These Regulations came into force on 1 January 1993. The Regulations apply to the manual handling of loads and seek to prevent injury, not only to the back, but to any part of the body. Account is taken of physical properties of loads which may affect grip or cause injury by slipping, roughness, sharp edges, extremes of temperature.

The Regulations require that where there is the possibility of risk to employees from the manual handling of loads, the employer should take the following measures, in this order:

1) avoid hazardous manual handling operations so far as is reasonably practicable;
2) assess any hazardous manual handling operations that cannot be avoided; and
3) reduce the risk of injury so far as is reasonably practicable.

Steps taken to avoid manual handling or reduce the risk of injury must be regularly checked to see if they are effective.

It is a requirement of the HSW Act and the Management of Health and Safety at Work Regulations 1992 that employers provide their employees with health and safety information and training. This should include specific information and training on manual handling, injury risk and prevention, as part of the steps to reduce risks required by these Regulations.

Although the Regulations do not set out specific requirements such as weight limits they do give numerical guidelines to assist with assessment. Guidelines for lifting and lowering are shown in Fig. 2.1 This shows guideline figures taking into consideration

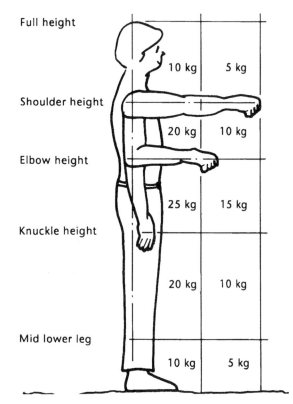

Figure 2.1 Lifting and lowering loads.

vertical and horizontal position of the hands as they move during the handling operation. e.g. 10 kg if lifted to shoulder height at arm's length or 5 kg if lifted to full height at arm's length. This assumes that the load can be easily grasped, with a good body position and in reasonable working conditions. If the hands enter more than one of the boxes during lifting, the smallest weight figure should be used.

Good handling techniques

The development of good handling technique is no substitute for the risk reduction steps already outlined but is an important addition which requires training and practice. The following should form the basic lifting operation.

Stop and think: plan the lift. Organise the work to minimise the amount of lifting necessary. Know where you are going to place the load. Use mechanical assistance if possible. Get help if load is too heavy. Make sure your path is clear. Don't let the load obstruct your view. For a long lift, i.e. from floor to shoulder height, consider a rest mid-way on a bench in

order to adjust your grip. Alternatively lift from floor to knee then from knee to carrying position – reverse this method when setting the load down

Place your feet: keep your feet apart to give a balanced and stable base for lifting (see Fig. 2.2). Your leading leg should be as far forward as is comfortable.

Adopt a good posture: keep your back straight and upright. Bend your knees and let your legs do the work (see Fig. 2.3). Keep your shoulders level and facing the same direction as your hips, i.e. don't twist your body.

Get a firm grip: lean forward a little over the load to get a good grip. Try to keep your arms within the boundary formed by the legs. Balance the load using both hands. A hook grip is less tiring than keeping the fingers straight. Wear gloves if the surface is rough or has sharp edges. Take great care if load is wrapped or slippery in any way.

Don't jerk: carry out the lifting operation smoothly, keeping control of the load.
Move the feet: don't twist your body if you turn to the side.

Keep close to the load: if the load is not close when lifting, try sliding it towards you. Keep the load close to your body for as long as possible. Keep the heaviest side of the load next to your body (see Fig. 2.4).

Put down: Putting down the load is the exact reverse of your lifting procedure. If precise positioning of the load is required, put it down, then slide it to the desired position.

Whenever possible make use of mechanical assistance involving the use of handling aids.

Figure 2.2 Placing the feet.

Figure 2.3 Adopting a good posture.

Although an element of manual handling is still present, body forces are applied more efficiently reducing the risk of injury. Levers can be used which lessen the body force required and also remove fingers from a potentially dangerous area. Hand- or power- operated hoists can be used to support a load and leave the operator free to control its positioning. A trolley, sack truck or roller conveyer can reduce the effort required to move a load horizontally while chutes using gravity can be used from one height to the next. Hand-held hooks and suction pads can be used where loads are difficult to grasp.

As a general rule, loads over 20 kg need the assistance of lifting gear.

Transporting loads

Light loads can be transported with the aid of a variety of hand trucks and trailers largely depending on the shape of the articles.

Hand trucks (Fig. 2.5): with two wheels having a capacity of around 250 kg can be used to transport sacks and boxes or special types are available for transporting oil drums, gas cylinders and similar items.

Flat trolley or trailers (Fig. 2.6): with four wheels are more stable and capable of moving heavier loads with a typical capacity of between 500 and 800 kg depending on the design. These are used where uniform shapes such as crates and boxes have to be transported. Where this can be done safely, the load can be stacked, but care must be taken not to stack too high or the load may become unstable and dangerous.

Palletised loads can be easily moved with the aid of hand pallet trucks as shown in Fig. 2.7. The forks are simply pushed under the pallet, a hydraulic lifting system raises the pallet and the easy-going wheels enable the load to be placed where required. These trucks have a loading capacity of up to 3000 kg.

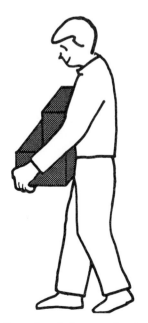

Figure 2.4 Keep load close to body.

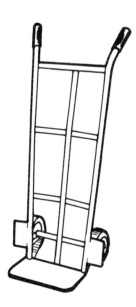

Figure 2.5 Hand truck.

Figure 2.6 Flat trailer.

There are a number of **do's** and **do not's** to observe regarding safe stacking either in storage areas or during transportation:

Do

■ always wedge objects that can roll, e.g. drums, tubes, bars;
■ keep heavy objects near floor level;
■ regularly inspect pallets, containers, racks for damage;
■ stack palletised goods vertically on a level floor so they won't overbalance.
■ 'key' stacked packages of uniform size like a brick wall so that no tier is independent of another;
■ use a properly constructed rack secured to a wall or floor.

Figure 2.7 Hand pallet truck.

Do not

■ allow items to stick out from racks or bins into the gangway;
■ climb racks to reach upper shelves – use a ladder or steps;
■ lean heavy objects against structural walls;
■ de-stack by climbing up and throwing down from the top or pulling out from the bottom;
■ exceed the safe working loading of racks, shelves or floors.

Power lifting

Where loads are too heavy to be manually lifted, some form of lifting equipment is required.

Lifting equipment in use should be:

■ of good construction, sound material, adequate strength, free from patent defect and be properly maintained;
■ designed so as to be safe when used;
■ proof tested by a competent person before being placed into service for the first time and after repair;
■ thoroughly examined by a competent person after proof testing;
■ certified on the correct test certificate including proof load and the safe working load (SWL);
■ marked with its SWL and with a means of identification to relate to its certificate;
■ thoroughly examined by a competent person within the periods specified in the Acts (see Table 2.1) and a record kept in a register.

Lifting equipment can be classified according to its power source: manual, electrical, pneumatic, hydraulic, or petrol or diesel.

Manual

Muscle power is restricted to the manual effort required to operate portable lifting equipment. An example of this is the chain block shown in Fig. 2.8 capable of lifting loads up to around 5000 kg. The mechanical advantage is obtained through the geared block thereby reducing the manual effort required in operating the chain.

Table 2.1

Type of equipment			
Lifting machines, cranes, pulley blocks	Chains, wire ropes slings, rings, hooks, shackles	Wire ropes	Fibre ropes
Periodic examination			
After testing and every 14 months	After proof testing and every six months	On cranes, inspect in position every four weeks	Every six months
Proof load testing			
Before being taken into use and after repair	Before being taken into use and after repair	Sample of rope tested to destruction	Not applicable

Small hydraulically operated portable cranes (Fig. 2.9) are available which have an adjustable jib with a lifting capacity typically from 350 to 550 kg on the smaller models up to 1700–2500 kg on the larger models. The smaller figure is the load capable of being lifted with the jib in its most extended position. In this case the mechanical advantage is obtained through the hydraulic system again reducing the manual effort.

Figure 2.8 Chain block.

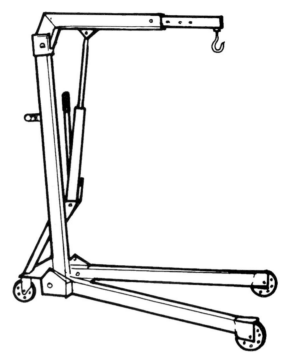

Figure 2.9 Portable crane.

Electrical

Electric motors are used to power chain blocks, hoists such as vehicle lifting platforms and large overhead gantry cranes capable of lifting substantial loads.

Pneumatic

Air chain hoists are available operated by air motors driven by compressed air at either 4 or 6.3 bars. Safe working loads range from 0.25 to 50 tonnes. These may also be linked to air-driven trolleys where hoists are used in conjunction with an overhead gantry.

Exhausting air from a system is commonly used in lifting. Powered vacuum lifters are used to lift sheet material, plastics sacks and cartons in a range of industries.

Hydraulic

Hydraulic systems use pressurised liquid, usually oil. This can be in the form of a hydraulic cylinder moving a heavy load over a short distance or through a pump to a hydraulic motor linked to a mechanical means of movement, e.g. cables or screws.

Petrol or diesel

This power source is not used in a factory environment but is common for on-site work in hoists used for raising and lowering materials and equipment. They are totally portable and independent.

Lifting gear accessories

Hooks: are forged from hardened and tempered high-tensile steel and are rated according to their safe working load (SWL). A range of types are available including those with a safety catch as shown in Fig. 2.10. Form and dimension are set out in BS2903:1980 which states that all hooks shall be free from patent defect and shall be cleanly forged in one piece. After manufacture and heat treatment each hook shall be subjected to a proof load. After proof testing each hook is stamped to allow identification with the manufacturer's certificate of test and examination.

Slings: are manufactured from a number of materials generally man-made fibre, wire rope and chain.

Belt slings made from a high tenacity polyester webbing are commonly used. One manufacturer colour codes the slings and weaves a stripe into the webbing to identify the SWL, for example:

1 tonne SWL – one stripe woven into a blue webbing
2 tonne SWL – two stripes woven into a green webbing
3 tonne SWL – three stripes woven into a yellow webbing
4 tonne SWL – four stripes woven into a orange webbing.

Figure 2.10 Hook.

Belt slings are available in standard lengths from 2 to 6 metres and in widths from 50 to 200 mm. They are available up to 12 tonnes SWL and may be endless, fitted with two soft eye lifting loops for general-purpose lifting or lightweight metal fittings for improved protection from pressure at the lifting point (see Fig. 2.11).

Figure 2.11 Belt sling.

If a lift is vertical (i.e. the lifting hook on one end and the load vertically below on the other end), the total load that may be lifted is that marked on the sling. However, if the method of lifting differs, the working load limit (WLL) will alter as shown in Table 2.2. The figures relate to a sling with a 1 tonne (1000 kg) SWL.

As shown in Table 2.2, a sling with a SWL of 1 tonne used in a basket hitch lift, has a working load limit (WLL) of 2 tonnes.

Table 2.2

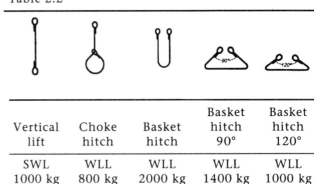

Vertical lift	Choke hitch	Basket hitch	Basket hitch 90°	Basket hitch 120°
SWL 1000 kg	WLL 800 kg	WLL 2000 kg	WLL 1400 kg	WLL 1000 kg

Wire rope and chain slings may be endless, single leg or have two, three or four legs with various eyes and fittings at the ends, such as shackles, links, rings and hooks. Figure 2.12(a) shows a three-leg wire rope sling and Fig. 2.12(b) a two-leg chain sling.

Slings are proof tested, identified and marked with the SWL and provided with a dated test certificate.

It is against the law to use lifting equipment

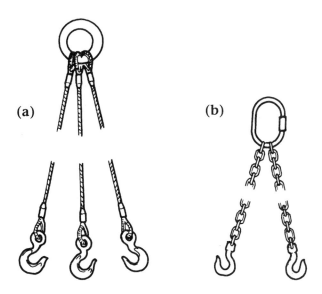

Figures 2.12 (a) Three-leg chain sling. **(b)** Two-leg chain sling.

Table 2.3

Size	Single leg	Two leg 0°–90°	90°–120°	Three leg	Four leg 0°–90°	90°–120°
mm 10	kg 3200	kg 4500	kg 3200	kg 6700	kg 6700	kg 4800

Table 2.4

Rope dia.	Single leg	Two leg 0°–90°	90°–120°	Three leg	Four leg 0°–90°	90°–120°
mm 10	kg 1200	kg 2200	kg 1600	kg 3200	kg 3200	kg 2300

or accessories for a load greater than the SWL except in the case of a multi-legged sling, where a table showing the load limits at different angles of legs must be posted in prominent positions in the factory. However, it is now becoming common to mark a multi-legged sling with one SWL, e.g. SWL Xt 0°–90°. This means that the sling has a SWL rating corresponding to an angle of 90° and this applies to all angles between 0° and 90° between the legs.

In the case of a four-legged sling the angle is taken between diagonally opposite legs and in a three-legged sling it is the angle between two adjacent legs.

The working load limit (WLL) of wire rope and chain slings vary with the number of legs and the angle of the legs in operation.

Table 2.3 shows the working load limits for a variety of chain slings whose links are made from 10 mm diameter material and operating at various angles (see BS6304:1982).

Table 2.4 shows the working load limits for a variety of wire rope slings made from 10 mm diameter wire and operating at various angles (see BS6210:1983).

Collar eyebolt (Fig. 2.13): Some larger components are fitted with a drilled and tapped hole to accept an eyebolt to simplify lifting. Collar eyebolts, manufactured to BS4278:1984 are available with threads from 12 to 45 mm with larger sizes available to order. Imperial

threads are available for replacement on older equipment. Typical SWL of an eyebolt having a 45 mm thread is 8 tonnes.

Care must be taken to avoid mismatching threads i.e. screwing a metric eyebolt into an imperial threaded hole and vice versa. When fitted, the eyebolt should be screwed against the face of the collar which should sit evenly on the contacting surface. If a single eyebolt is used for lifting a load which is liable to revolve or twist, a swivel type hook should be used to prevent the eyebolt unscrewing.

Collar eyebolts may be used up to the SWL

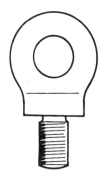

Figure 2.13 Collar eyebolt.

Figure 2.14 Eyebolt with link.

Table 2.5

Single eyebolt vertical SWL	Maximum load (*W*) lifted by pair of eyebolts when angle (*A*) between sling is		
	up to 30°	30° to 60°	60° to 90°
2000 kg	2500 kg	1600 kg	1000 kg

Table 2.6

Single eyebolt vertical SWL	Maximum load (*W*) lifted by pair of eyebolts with links when angle (*A*) between sling is		
	up to 30°	30° to 60°	60° to 90°
2000 kg	4000 kg	3200kg	2500 kg

for axial lifting only. Eyebolts with a link (Fig. 2.14), offer considerable advantages over collar eyebolts, when loading needs to be applied at angles to the axis (Fig. 2.15). Their SWL is relatively greater than those of the plain collar eyebolt and the load can be applied at any angle.

The maximum recommended working load at various angles (as indicated in Fig. 2.15) is set out in Table 2.5 for collar eyebolts and in Table 2.6 for eyebolts with a link.

Shackles (Fig. 2.16): made from alloy steel and manufactured to BS3551:1962 are used in conjunction with lifting equipment and accessories. There are four types of pin which can be fitted but the most common is the screwed pin with eye and collar as shown. The range of pin diameters vary from 16 mm up to 108 mm. A range of pin diameters and their SWL is shown in Table 2.7.

Chains: with welded links in alloy steel are manufactured in accordance with BS3113: 1985. The British Standard covers the designa-

tion of size, material used and its heat treatment and dimensions, e.g. material diameter, welds and dimensions of links. The welds should show no fissures, notches or similar faults and the finished condition should be clean and free from any coating other than rust preventative.

All chain is tested, proof loaded and marked to comply with the BS requirements and issued with a certificate of test and examination together with particulars of the heat treatment which the chain was subjected to during manufacture.

Special-purpose equipment: where production items are regularly lifted, lifting and spreader beams can be used, specially designed for the purpose. Standard lifting beams (Fig. 2.17a) or spreader beams (Fig. 2.17b) are used to give a vertical lift where the lifting points are too far apart to use slings.

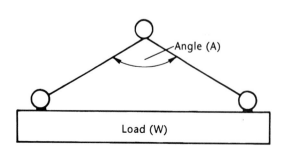

Figure 2.15 Loading at angle using eyebolt with link.

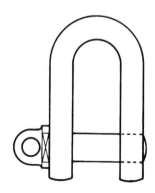

Figure 2.16 Shackle.

Table 2.7

Diameter of pin (mm)	16	25	32	38	48	60	70	83	108
SWL tonnes	1.1	4.5	7.5	10.5	16.8	27	35	50	80

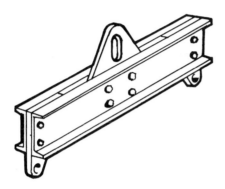

Figure 2.17(a) Lifting beam.

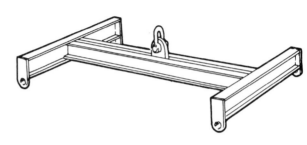

Figure 2.17(b) Spreader beam.

For lifting items such as steel drums and pipes, chain slings can be fitted with special hooks. Figure 2.18 shows pipe hooks, which are used in pairs.

Where large plates are to be lifted, plate clamps (Fig. 2.19) are available which are used to vertically lift plate up to 130 mm thick and having a SWL of 30 tonnes.

Safe lifting

There are a range of safety rules to observe when carrying out any lifting operation.

■ use only equipment marked with its safe working load (SWL) for which there is a current test certificate;
■ never exceed the safe working load (SWL) of machines or tackle. (Some machines are fitted with a load limiting device which cuts off the power supply if an overload condition develops);
■ never lift a load if you doubt its weight or the adequacy of the equipment;
■ before lifting an unbalanced load find its centre of gravity. Raise it a little off the ground and pause – there will be little harm if it drops;

■ never use makeshift, damaged or badly worn equipment e.g. chains shortened with knots, kinked or twisted wire ropes, frayed or rotted fibre ropes;
■ use suitable packing to protect slings from damage by sharp edges of loads;
■ do not allow tackle to be damaged by being dropped or dragged from under a load;
■ avoid snatching a load;
■ make sure that people or loads cant fall from a high level;
■ never transport loads over the heads of people or walk under a suspended load;
■ never leave a load hanging unnecessarily for any period of time.

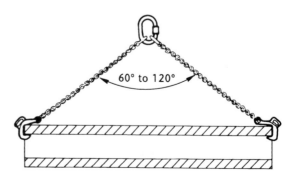

Figure 2.18 Pipe hooks in use.

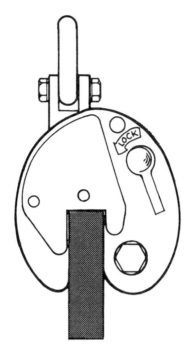

Figure 2.19 Plate clamp for vertical lift.

Knots

When using natural or man-made fibre ropes, it is necessary to have a good knowledge of knots in order to safely join them or safely secure loads.

Remember – never knot wire ropes or chains. Some of the commonly used knots are described and shown in Fig. 2.20.

Reef knot – a common and widely used knot for joining ropes of equal thickness or two ends of the same rope.

Sheetbend – used to join ropes of unequal thickness.

Clove hitch – used for tying a rope to a bar or round object. This knot is most satisfactory when the force is equal in both directions. Where the force is on one side, use the round turn and two half hitches.

Bowline – used to make a loop which will not slip.

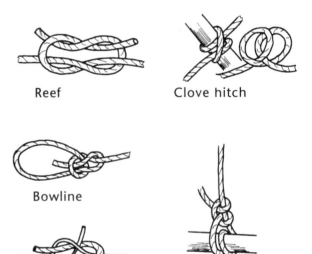

Reef   Clove hitch

Bowline

Sheetbend   Round turn and two half hitches

Figure 2.20 Knots.

3

Drawings, specifications and data

The technical drawing is the means by which designers communicate their requirements. A technical drawing is a pictorial representation of a component with dimensions and other data added. Together with various specifications, the drawing sets out a detailed description of the part which is to be made. The arrangement of dimensions shown on the drawing determine the shape and together with various characteristics, will have an influence on how the part is made, e.g. whether the part is to be produced from a casting or forging, whether it is a fabricated or welded structure or machined from solid material.

The material specification will include the type of material required together with any standard parts such as nuts, bolts, washers, etc.

Types of tools to be used in the manufacture of the part may also be determined, e.g. if the drawing specifies an 8 mm reamed hole. There may be a requirement for special equipment necessary for installation or assembly.

Specifications such as tolerances, surface finish, heat treatment or protective finishes such as painting or plating all influence the sequence of operations for manufacture.

You can see from this that a technical drawing and its associated specification, communicate a large amount of information.

Drawings are a fundamental part of engineering and are used from the moment design work starts on a new product, through all its stages, to the final finished product.

Standardisation

In order to create the necessary uniformity in communicating technical information, a system of standardisation has to be adopted. The standardisation can be the information contained on a drawing, the layout of the drawing, the standardisation of sizes or of parts.

This standardisation can be adopted within an individual firm or group of firms, or within an industry, e.g. construction, shipbuilding, etc., nationally by adopting British Standards or internationally through ISO.

Adopting a system of standard specification, practices and design results in a number of advantages:

- reduction in design costs
- reduction in cost of product
- redundant items and sizes are eliminated
- designs are more efficient
- level of interchangeability is increased
- mass-production techniques can be adopted
- purchasing is simplified
- control of quality is enhanced
- spares can be easily obtained
- costing can be simplified
- overheads are reduced.

Except for written notes, technical drawings have no language barriers. They provide the universal language for design, for the craftsman and the technician in manufacture, assembly and maintenance, for the sales team as an aid to selling and for the customer before buying or indeed servicing after purchase.

Communicating technical information

Many different methods are used to communicate technical information. The method chosen will depend on how much information has to be dispensed and its complexity. Whichever method is chosen the all-important factor is that the information is simply presented, easy to understand and unambiguous.

Technical drawings

Technical drawings can vary from thumbnail sketches to illustrate a particular piece of information through pictorial drawings in isometric or oblique projection to major detail and assembly drawings. These are covered later in this chapter.

Diagrams

Diagrams are used to explain rather than represent actual appearances. For example, an electrical circuit diagram shows the relationship of all parts and connections in a circuit represented by lines and labelled blocks without indicating the appearance of each part.

Figure 3.1 is a diagram to explain the route of oil circulation in a car engine but does not go into detail of the engine itself.

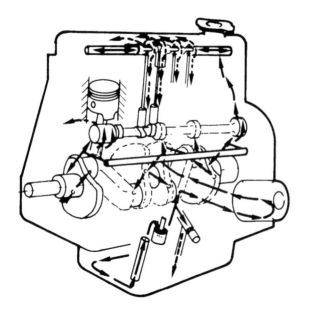

Figure 3.1 Engine oil circulation diagram.

Exploded views

Exploded views are used where it is necessary to show the arrangement of an assembly in three dimensions. They are used for assembly purposes and in service or repair manuals where reference numbers of the parts and the way in which they fit together is shown. Figure 3.2 shows an exploded view of a hand riveting tool, listing the spare parts, their identification numbers and their relative position within the finished product.

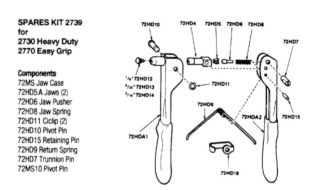

Figure 3.2 Exploded view of hand riveting tool.

Operation sheet

The purpose of an operation sheet is to set out the most economic sequence of operations required to produce a finished object or process from the raw material. Although the main purpose of operation sheets is to set out the sequence of operations, they also serve a number of other very important functions:

■ they determine the size and amount of material required. From this information, the material can be ordered in advance and appropriate material stock levels can be maintained;
■ any tooling such as jigs, fixtures and gauges can be ordered or manufactured in advance so that it will be available when required;
■ knowing the machines or plant which are to be used enables machine-loading charts to be updated so that delivery dates to customers will be realistic;
■ the sequence of operations listed will enable work to be progressed through the factory in an efficient manner;

■ the inclusion of estimated times for manufacture on the operation sheet enables a cost of manufacture and hence selling price to be established.

Figure 3.3 shows a fitted bolt with a hexagon head which is to be produced in one operation on a capstan lathe. The sequence of operations is shown in operation sheet 1.

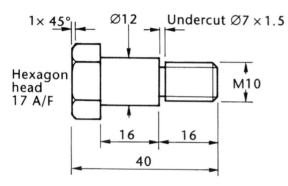

Figure 3.3 Hexagon headed fitted bolt.

Data sheets

Data sheets are available where a range of information is available and where you can obtain the information required for your application. Table 3.1 shows data relating to a range of tin–lead base solders and their use. Table 4.3 on page 62 shows a data sheet from BS4500 on tolerance grades.

There are many examples of data sheets some of which are made up as wall charts for easy reference.

Tables

Tables are used to show a range of options or sizes available and help with a choice for a particular set of circumstances. Table 3.2 shows the selection of a hacksaw blade with the correct number of teeth for the task, given the type and thickness of material to be cut.

Operation sheet 1. Material: 17 A/F × 42 mm. M.S. hexagon bar. Name: fitted bolt

Op. no.	Machine	Operation	Tooling	Position
1	Capstan lathe	Feed to stop	Adjustable stop	Turret 1
		Turn 10 mm thread diameter	Roller box tool	Turret 2
		Turn 12 mm diameter	Roller box tool	Turret 3
		Face end and chamfer	Face and chamfer tool	FTP
		Form undercut	Undercut tool	FTP
		Die thread	Self-opening diehead	Turret 4
		Chamfer head	Chamfer tool	FTP
		Part off	Parting-off tool	RTP

FTP – front toolpost; RTP – rear toolpost.

Table 3.1

BS219	Composition (%)		Melting range (°C)	Uses
	Sn	Pb		
A	64	36	183–185	Mass soldering of printed circuits
K	60	40	183–188	General soldering
F	50	50	183–220	Coarse tinman's solder
G	40	60	183–234	Coating and pre-tinning
J	30	70	183–255	Electrical cable conductors
W	15	85	227–288	Electric lamp bases

Sn – chemical symbol for tin; Pb – chemical symbol for lead.

Table 3.2

Workpiece material	Workpiece thickness in mm			
	3	6	10	13
Aluminium Free machining low carbon steels	32	24	18	14
Brass Copper Low and medium carbon steels	32		24	18 14
High carbon steels Bronze Alloy steels Stainless steels Cast iron	32	24		18

Tables are also available setting out tapping and clearance drill sizes for a range of thread sizes. Table 3.3 shows such a table for ISO metric coarse threads. These are also available as wall charts for easy reference.

Table 3.3

ISO metric coarse thread			
Diameter (mm)	Pitch (mm)	Tapping drill (mm)	Clearance drill (mm)
2.0	0.40	1.60	2.05
2.5	0.45	2.05	2.60
3.0	0.50	2.50	3.10
4.0	0.70	3.30	4.10
5.0	0.80	4.20	5.10
6.0	1.00	5.00	6.10
8.0	1.25	6.80	8.20
10.0	1.50	8.50	10.20
12.0	1.75	10.20	12.20

Graphs

Graphs are a useful means of communication where there is a relationship between sets of technical data. Figure 3.4 shows the relationship between tempering temperature and hardness of a particular type of steel. From the graph it can be seen that to achieve a final hardness of 60 RC, a part made from this steel would be tempered at 200°C.

Storing technical information

Storing large amounts of original drawings in paper form is very bulky and takes up a lot of space. Original drawings can be stored on microfilm by photographically reducing them on to film which is then mounted and suitably filed. When required they can be re-enlarged and printed for use on the shop floor. Care must be taken when preparing drawings which are to be microfilmed otherwise some detail may be lost in the reducing process.

Libraries and spares departments now carry complete documents or lists of spares on microfiche systems. The information is stored

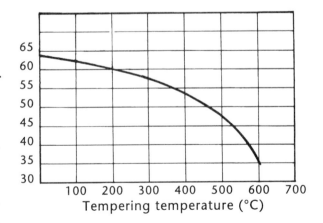

Figure 3.4 Temperature/hardness graph for tempering.

on a large film which is divided into a number of frames or grids identified by letters along one edge and numbers along the other. The information is viewed on a screen by moving to the required grid reference, e.g. B12.

A great deal of technical drawing and design is now being done on computer-aided draughting or design (CAD) systems where the computer is capable of generating and storing much technical information. This information can be stored in the computer itself on the hard disk or on separate microdisks for retrieval at any time. Any modification can be easily carried out. Information can be easily retrieved and viewed on a visual display unit and when required can be downloaded to a printer or plotter. Furthermore, the information can also be accessed by others in the same factory through computer networks or sent using a modem, through the telephone system to other factories in other parts of the country.

Video tapes are increasingly being used in technical, sales and education sectors. Again the information contained can be easily accessed and sent anywhere in the country without the need and expense of actual machines or equipment being present.

Interpreting drawings

Presenting information in the form of a drawing is the most appropriate method of communication. Drawings provide a clear and concise method of conveying information to other individuals and other departments. They give a permanent record of the product and enable identical parts to be made at any time and if modifications are required these can be easily incorporated and recorded. Finally as has already been stated, they do not represent a language barrier.

When an initial design is decided, it is broken down into a series of units for assembly or sub-assembly, which are further broken down into individual parts referred to as detail drawings.

Assembly drawing

An assembly drawing shows the exact relative position of all the parts and how they fit together. Dimensions of individual parts are not shown although indication of overall dimensions relative to, and important for the final assembly may be included. All individual parts are identified by an item number. A parts list is shown giving information such as: item number, part description, part number and quantity required. The parts list would include details of bought-out items such as nuts, bolts, screws, bearings, electric motors, valves' etc.

As the name implies these drawings are used by engineers to fit together, or assemble, all the parts as the final product and by maintenance engineers during servicing, repair or replacement of worn parts. An example of an assembly drawing is shown in Fig. 3.5.

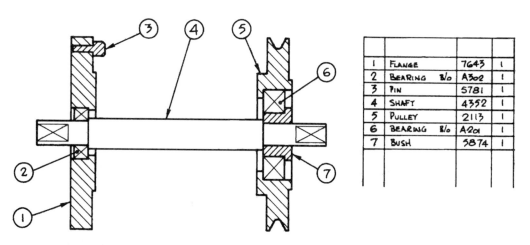

1	FLANGE		7643	1
2	BEARING	B/o	A302	1
3	PIN		5781	1
4	SHAFT		4352	1
5	PULLEY		2113	1
6	BEARING	B/o	A201	1
7	BUSH		5874	1

Figure 3.5 Assembly drawing.

Detail drawing

A detail drawing is done to give production departments all the information required to manufacture the part. Every part required in an assembly, except standard or bought-out items will have its own detail drawing.

The information given on a detail drawing must be sufficient to describe the part completely and would include:

- dimensions
- tolerances
- material specification
- surface finish
- heat treatment
- protective treatment
- tool references
- any notes necessary for clarity

All detail drawings should have a title block usually located at the bottom right-hand corner. For ease of reference and filing of its drawings, a company should standardise the position and content of a title block and this is pre-printed on the drawing sheet. The information contained will vary from one company to another but would be likely to include most of the following:

- name of the company
- drawing number
- title
- scale
- date
- name of draughtsman/woman
- unit of measurement
- type of projection
- general tolerance notes
- material specification

An example of a title block is shown in Fig. 3.6.

Block plan

Block plans are used in the construction industry to identify a site and locate the outlines of a construction in relation to a town plan or other context.

Location drawing

Location drawings are used in the construction industry to locate sites, structures, buildings, elements, assemblies or components.

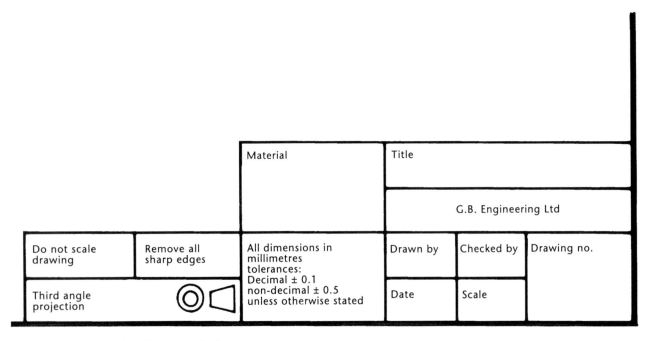

Figure 3.6 Example of a title block.

Projection

The two common methods used to represent a three-dimensional object on a flat piece of paper are orthographic projection and pictorial projection.

Orthographic projection: this is the most frequently used method of presenting information on a detail drawing. It is usually the simplest and quickest way of giving an accurate representation, and dimensioning is straightforward. It is only necessary to draw those views that are essential and it is simple to include sections and hidden detail.

Orthographic projection may be first angle or third angle and the system used on a drawing should be shown by the appropriate symbol.

An example of first angle projection is shown in Fig. 3.7 each view showing what would be seen by looking on the far side of an adjacent view.

An example of third angle projection is shown in Fig. 3.8 each view showing what would be seen by looking on the near side of an adjacent view.

Pictorial projection: this is a method of showing, on a single view, a three-dimensional picture of an object. It is therefore easier to visualise and is useful for making rough sketches to show someone what you require.

Two kinds of pictorial projection are used: isometric projection and oblique projection.

Isometric projection – in isometric projection, vertical lines are shown vertical but horizontal lines are drawn at 30° to the horizontal on each side of the vertical.

This is shown by the rectangular box in Fig. 3.9 and in Figs 3.7 and 3.8.

With this method, a circle on any one of the three faces is drawn as an ellipse. In its simplest form, all measurements in orthographic views may be scaled directly onto the isometric view.

Figures 3.7 and 3.8 show an isometric view and the corresponding orthographic views.

Oblique projection – the main difference between isometric and oblique projection is that with oblique projection one edge is horizontal. Vertical edges are still vertical which means two axes are at right angles to each other. This means that one face can be drawn

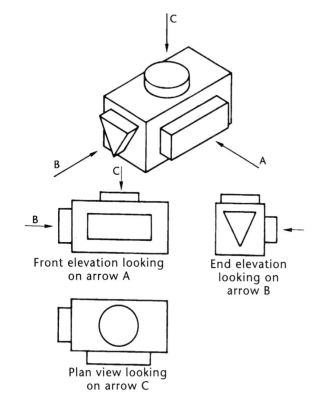

Figure 3.7 First angle projection.

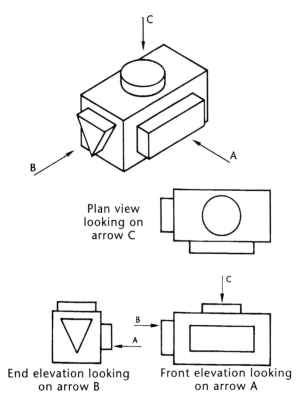

Figure 3.8 Third angle projection.

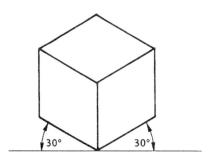

Figure 3.9 Isometric projection.

as its true shape. The third edge can be drawn at any angle but is usually 30° or 45°. This is shown by the rectangular box in Fig. 3.10.

To avoid a distorted view, the dimensions along the receding edges are drawn half full size.

The rule to follow when drawing oblique projection is to chose the face with most detail as the front face so it can be drawn as its true shape e.g. a circle remains a circle and can be drawn using a pair of compasses. This will make use of the advantage that oblique has over isometric and will show the object with minimum distortion.

Figure 3.11 shows an oblique view and two corresponding orthographic views.

Sectional views

A part with little internal detail can be satisfactorily represented by orthographic projection. However, where the internal detail is complicated and are represented by hidden detail lines, the result may be confusing and difficult to interpret correctly. In such cases, the interior detail can be exposed by 'cutting away' the outside and showing the inside as full lines instead of hidden detail. The view thus drawn is known as a sectional view.

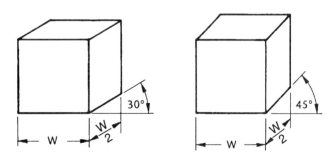

Figure 3.10 Oblique projection.

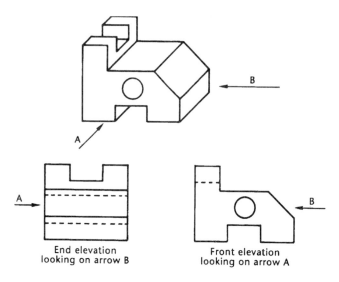

End elevation
looking on arrow B

Front elevation
looking on arrow A

Figure 3.11 Oblique and corresponding orthographic views.

There are a range of sectional views which can be done and these are shown in BS308 Engineering Drawing Practice. Figure 3.12 shows a single plane sectional view, i.e. an imaginary slice taken straight through the object called the 'cutting plane'. The cutting plane is shown by long chain lines, thickened at the ends and labelled by capital letters. The direction of viewing is shown by arrows resting on the thickened lines. The area which is sectioned is hatched by thin lines at 45° to the part profile.

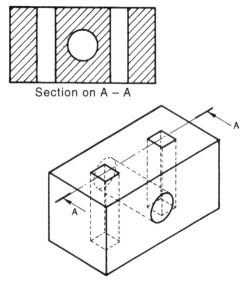

Section on A – A

Figure 3.12 Sectional view.

Standard conventions

Because an engineering drawing is the means of communication it is important that all drawings are uniform in their layout and content. BS308: Engineering Drawing Practice sets out to standardise conventions used in engineering drawings which include: layout, lines, systems of projection, sections, conventional representation and dimensions.

Layout of drawings

Layouts should use the 'A' series of sheet sizes A4, A3, A2, A1 and A0 the dimensions of which are shown in Table 3.4. The drawing area and title block should be within a frame border.

Table 3.4

Name	Size in mm
A4	210 × 297
A3	297 × 420
A2	420 × 594
A1	594 × 841
A0	841 × 1189

Lines and line work

All lines should be uniformly black, dense and bold. Types of line are shown in Table 3.5.

Table 3.5

Type of line	Description of line	Application
———	Thick, continuous	Visible outlines and edges
———	Thin, continuous	Dimension and leader lines Projection lines Hatching
- - - - - -	Thin, short dashes	Hidden outlines and edges
—·—·—·—	Thin, chain	Centre lines
⌐·—·⌐	Chain, (thick at ends and at changes of direction, thin elsewhere)	Cutting planes of sections

Systems of projection

The system of projection used on a drawing should be shown by the appropriate symbol as shown in Fig. 3.13.

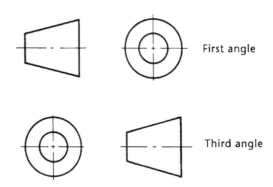

Figure 3.13 Symbols for system of projection.

Conventional representation

Standard parts which are likely to be drawn many times results in unnecessary waste of time and therefore cost. Conventional representation is used to show common engineering parts as simple diagrammatic representations. These are shown in BS308, a selection of which is shown in Fig 3.14.

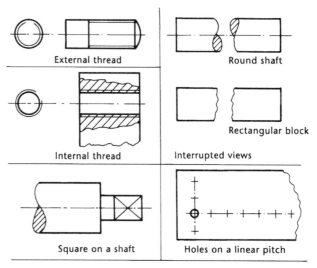

Figure 3.14 Conventional representation.

Dimensioning

All dimensions necessary for the manufacture of a part should be shown on the drawing and should appear once only. It should not be necessary for a dimension to be calculated or for the drawing to be scaled.

Dimensions should be placed outside the outline of the view wherever possible. Projection lines are drawn from points or lines on the view and the dimension line placed between them. Dimension lines and projection lines are thin continuous lines. There should be a small gap between the outline and the start of the projection line and the projection line should extend for a short distance beyond the dimension line. The dimension line has an arrow-head at each end which just touches the projection line.

Dimension lines should not cross other lines unless this is unavoidable.

Leaders are used to show where dimensions or notes apply and are thin continuous lines ending in arrow-heads touching a line or as dots within the outline of the object. These principles are shown in Fig. 3.15.

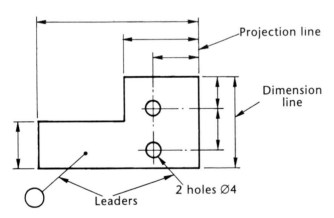

Figure 3.15 Principles of dimensioning.

Difficulties can arise if the dimensioning of a part is not done correctly, which may lead to errors of the finished part.

In Fig. 3.16(a) the part is dimensioned from the datum in sequence i.e. cumulatively. Each dimension is subject to the tolerance of ±0.15 so that a cumulative error of ±0.45 in the overall length is possible.

By dimensioning the part as shown in Fig. 3.16(b) the cumulative error is avoided.

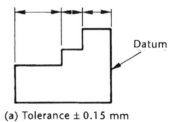

(a) Tolerance ± 0.15 mm

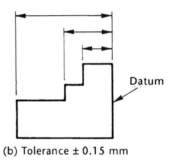

(b) Tolerance ± 0.15 mm

Figure 3.16 Dimensioning to avoid cumulative error.

Difficulties also arise if a part is dimensioned from more than one datum. The shouldered pin shown in Fig. 3.17(a) has been dimensioned from each end, i.e. two separate datums. This could result in a possible maximum error of ±0.6 mm between the shoulders A and B.

If the dimension between shoulders A and B is important then it would have to be individually dimensioned relative to one datum and subject to a tolerance of ±0.2 mm as shown in Fig 3.17(b).

Difficulties may also arise where several parts are assembled and an important dimension has to be maintained. The datum chosen must reflect this important dimension. Figure 3.18a) shows an assembly where A is the important dimension. The chosen datum face is indicated. Thus the dimensioning of each individual part of the assembly reflects this datum as shown in Fig 3.18(b)–(d).

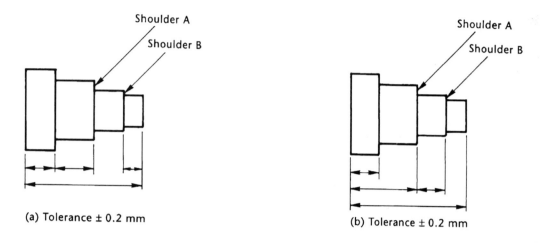

Figure 3.17 Dimensioning to maintain important size.

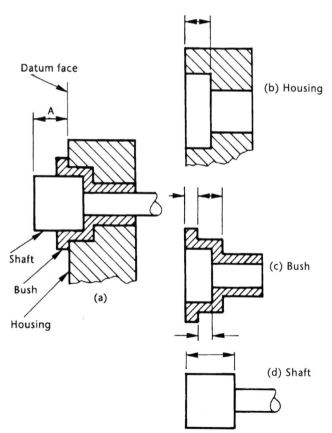

Figure 3.18 Dimensioning to maintain correct assembly.

Colour coding

Colours play an essential safety role in giving information for use in the prevention of accidents, for warning of health hazards, to identify contents of gas cylinders, pipeline and services, the identification and safe use of cables and components in electronic and electrical installations as well as the correct use of fire-fighting equipment.

The purpose of a system of safety colours and safety signs is to draw attention to objects and situations which affect or could affect health and safety. The use of a system of safety colours and safety signs does not replace the need for appropriate accident prevention measures.

Safety signs

British Standard BS5378:Part 1:1980 Safety Signs and Colours is concerned with a system for giving safety information which does not, in general, require the use of words. BS5499:Part 1:1990 extends the basic framework concerning safety colours and safety signs in BS5378 with regard to fire and on the inside front cover.

The safety colours, their meaning, and examples of their use are shown in Table 3.6. Examples of the shape and colour of the signs are shown in Figs 3.19–3.23 and on the inside front cover.

Red

Smoking prohibited

Figure 3.19 Prohibition – indicates certain behaviour is prohibited.

Red

Fire extinguisher

Figure 3.20 Fire equipment.

Table 3.6

Safety colour	Meaning	Examples of use
Red [white background colour with black symbols]	Stop Prohibition (**Don't** do)	Stop signs Emergency stops Prohibition signs
Red [white symbols and text]	Fire equipment	Position of fire equipment, alarms, hoses, extinguishers, etc.
Yellow [black symbols and text	Warning (risk of danger)	Indication of hazards (electrical, explosive, radiation, chemical, vehicle, etc.) Warning of threshold, low passages, obstacles
Green [white symbols and text]	Safe condition (the safe way)	Escape routes Emergency exits Emergency showers First-aid and rescue stations
Blue [white symbols and text]	Mandatory action (**MUST** do)	Obligation to wear personal safety equipment

Yellow

Caution
toxic hazard

Figure 3.21 Warning – indicates warning of possible hazard.

Blue

Head protection
must be worn

Figure 3.23 Mandatory – indicates specific course of action to be taken.

Green

First aid

Figure 3.22 Safe condition – conveys information about safe conditions.

Portable fire extinguishers

In most cases the entire body of a portable fire extinguisher is colour coded to indicate the medium contained. Colour coding by medium is intended to provide a means of rapid recognition of the type of extinguisher by trained persons at the time when the extinguisher is needed for use. The use of portable fire extinguishers was dealt with in Chapter 1.

Table 3.7 shows the extinguishing medium and the corresponding colour code which is also shown on inside back cover.

Table 3.7. Colour coding by medium

Extinguishing medium	Colour
Water	Signal red
Foam	Pale cream
Powder	French blue
Carbon dioxide (CO_2)	Black

Gas containers

Many thousands of people in industry use gas from cylinders in which the gas is contained at high pressure. All users should know and understand the properties of the gas they are using and the correct operating procedures for the equipment being used with the gas. Gas data and safety sheets are readily available from the supplier. As well as the container being marked with the gas it contains, colour is also used.

British Standard BS349:1973 'Identification of the Contents of Industrial Gas Containers sets out the colours used on cylinders to identify the gas contained and therefore relates to safety requirements. There is a range of gases available some of which may be either toxic, flammable or corrosive and the required safety precautions when transporting, handling, storing or using these must be observed.

Table 3.8 shows a range of industrial gases and the corresponding cylinder colour.

Table 3.8

Industrial gas	Cylinder colour
Acetylene	Maroon
Air	French grey
Argon	Peacock blue
Carbon dioxide (CO_2)	Black
Hydrogen	Signal red
Nitrogen	French grey with black band
Oxygen	Black
Propane	Signal red

Pipelines and services

It is essential to know what pipelines and service ducts contain, not only from the point of view of installation and maintenance but also from a safety aspect as would be the case in an emergency. If the pipe inadvertently fractures or in the event of fire, the contents would have to be known so that the most appropriate and safe action could be taken by the emergency services.

British Standard BS1710:1984 Identification of Pipelines and Services specifies the colours for the identification of pipes conveying fluids in above ground installations and on board ships. It also includes ducts for ventilation and conduits used for carrying electrical services. Where only the determination of the basic nature of the fluid is required the basic identification colour shall be applied by one of the following methods:

a) painted on the pipe over the whole length;
b) painted on the pipe as a band over a length of approximately 150 mm;
c) applied by wrapping an adhesive band or identification clip around the pipe of the basic identification colour over a length of about 150 mm.

Table 3.9 shows the basic identification colours and the corresponding contents.

Table 3.9

Pipe contents	Colour
Water	Green
Steam	Silver–grey
Oils – mineral, vegetable or animal	
Combustible liquids	Brown
Gases (except air)	Yellow ochre
Acids and alkalis	Violet
Air	Light blue
Other liquids	Black
Electrical services and ventilation ducts	Orange

Electrical wiring

IEE Regulations set out the colour identification used for electrical conductors.

The colour identification of cores of flexible cables and cords are:

- Live brown
- Neutral blue
- Earth green/yellow.

The colour identification of cores of non-flexible cables for fixed wiring are:

- Earth green/yellow
- Live a.c. single-phase circuit red
- Neutral a.c. single-phase circuit black
- Phase R of 3-phase a.c. circuit red
- Phase Y of 3-phase a.c. circuit yellow
- Phase B of 3-phase a.c. circuit blue.

4

Measurement and dimensional control

Some form of precise measurement is necessary if parts are to fit together as intended, no matter whether the parts were made by the same person, in the same factory or in factories in different parts of the world.

To achieve any degree of precision, the equipment used to measure must be precisely manufactured with reference to the same standard of length; that standard of length is the metre, which is defined in terms of the wavelength of a particular light. Measuring equipment relate to this standard and are then used to compare with the workpiece being measured.

Measuring instruments can be indicating or non-indicating.

In the case of indicating instruments, a readable measurement of the actual size of a workpiece can be made, e.g. in the use of a rule, micrometer or vernier caliper.

Non-indicating instruments merely indicate acceptance or non-acceptance of the workpiece, e.g. a plug gauge used to check a hole or gap gauge used to check a shaft. When checking a shaft with a gap gauge, the **GO** portion should pass over the shaft while the **NOT GO** portion should not pass over the shaft (see Fig. 4.1).

Similarly with a plug gauge used to check a hole – the **GO** portion should enter the hole, the **NOT GO** portion should not. Thus no readable measurement of the actual size is made, simply evidence of acceptance or non-acceptance. Using this type of equipment, workpieces can be quickly checked with little or no skill required, keeping costs to a minimum and is particularly suited to quantity production.

Standards

Standards are necessary in industry in order to ensure uniformity and to establish requirements of quality and accuracy. The adoption of standards eliminates the national waste of time and material involved in the production of an unnecessary variety of patterns and sizes of articles for one and the same purpose. Standards are desirable not only in the manufacture of articles but also for the instruments used to ensure the accuracy of these articles.

The word 'standard' can refer to a physical standard such as length or to a standard specification such as a paper standard. In the UK physical standards are maintained by the National Physical Laboratory (NPL). As the nation's standards laboratory, NPL provides the measurement standards and calibration facilities necessary to ensure that measurements in the UK may be carried out on a common basis and to the required accuracy. The

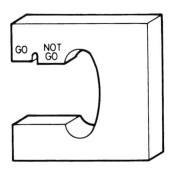

Figure 4.1 Solid gap gauge.

national primary standards, which constitute the basis of measurement in the UK, are maintained at NPL in strict accordance with internationally agreed recommendations based on the International System of Units (Système International d'Unités SI). The national primary standards are used to calibrate secondary standards and measuring equipment manufactured and used in industry.

The preparation of standard specifications in the UK is the responsibility of the British Standards Institution (BSI), whose main function is to draw up and promote the adoption of voluntary standards and codes of good practice by agreement among all interested parties, i.e. manufacturers and users. BSI plays a large and active part in the work of the International Organisation for Standardisation (ISO), which is responsible for the preparation of international recommendations. Standard specifications for materials used, heat treatment, dimensions and accuracy of a whole range of measuring equipment are provided by BSI in the form of British Standards. These British Standards are indicated by the letters 'BS' followed by a reference number, its year of introduction and the title to which it relates, e.g. BS6468:1984 Specification for Depth Micrometers.

Dimension standards

Standardisation of dimensional control is obtainable through BS4500 ISO Limits and Fits. This standard provides a comprehensive system of limits and fits for engineering. Although BS4500 refers to holes and shafts, the system applies to any feature such as squares, lengths, widths, depths relating to items such as keys, keyways, slots etc. This is covered on page 61.

Temperature

Length standards and measuring equipment being made of metal, expand as their temperature increases and the standard length for such equipment must relate to a standard temperature. The international standard temperature is 20°C. Where extremely accurate measurement is required, as in the case of manufacture or calibration of measuring equipment, this would be carried out under exacting conditions. Large organisations have what is known as a standards room, kept at a constant 20°C, with a controlled humidity level and dust-free atmosphere. (The SI unit for thermodynamic temperature is the kelvin, K. However, we use the more familiar celsius scale, in which, conveniently, one degree step of celsius is equivalent to one kelvin. However, the starting points are different: 0°C = 273 K. Therefore 20°C = 293 K.)

Advantages

Adopting a system of standards has a number of advantages. No two parts can ever be produced to the exact same size but if a system of dimension control is followed, then parts can be made, which when chosen at random, will fit together as intended. This is known as interchangeability and so parts made for immediate assembly will fit, as well as spare parts made at some later date. Being able to use standard 'off the shelf' parts such as nuts, bolts, bearings, material sizes, etc. leads to a reduced cost of the final product. Because standard parts are readily available, output can be increased. Collectively this gives a consistency of manufacture which results in improved quality.

SI units

Although we have only dealt with length units, there are seven basic units of the international system:

Quantity	Unit	Symbol
Mass	kilogram	kg
Length	metre	m
Time	second	s
Temperature	kelvin	K
Electric current	ampere	A
Luminous intensity	candela	cd
Amount of substance	mole	mol

these are the basic units from which all other units are derived. For instance velocity = distance (m) ÷ time (s) = m/s.

Quantities may be very large or very small and so a range of standard prefixes are used:

Factor	Prefix	Symbol
10^{12}	tera-	T
10^{9}	giga-	G
10^{6}	mega-	M
10^{3}	kilo-	k
10^{-3}	milli-	m
10^{-6}	micro-	µ
10^{-9}	nano-	n
10^{-12}	pico-	p

some common examples are: micrometre (µm), milliamp (mA), kilowatt (kW), megahertz (MHz).

Dimensional properties

An engineering drawing will give details of shape and size of the component. There are other properties however which although not shown on the drawing would be expected due to the production method used, e.g. if you turned and faced a bar on a lathe at the same setting, you would expect the face to be square to the turned diameter.

Properties include:

Length: the distance between two points, lines or surfaces and can be directly measured using a rule, micrometer or vernier caliper.

Straightness: an error in straightness of a feature may be stated as the distance separating two parallel straight lines between which the surface of the feature, in that position, will just lie. The workshop standard against which straightness of a line on a surface is compared, is the straight edge.

Flatness: an error in flatness of a feature may be stated as the distance separating two parallel planes between which the surface of the feature will just lie. Thus flatness is concerned with the complete area of a surface, whereas straightness is concerned with a line at a position on a surface; for example, lines AB, BC, CD, and DA in Fig. 4.2 may all be straight but the surface is not flat, it is twisted. The workshop standard against which the flatness of a surface is compared is the surface plate or table.

Parallelism: two surfaces which are a constant distance apart along their length or across their surface. This may be measured directly using a micrometer or vernier caliper, or using a surface table and dial test indicator.

Squareness: two surfaces are square when they are at right angles to each other. Thus the determination of squareness is one of angular measurement. There is no absolute standard for angular measurement in the same way as there is for linear measurement, since the requirement is simply to divide a circle into a number of equal parts. The workshop standard against which squareness is compared is the engineer's square.

Angle: when two surfaces are at an angle other than 90°, the angle between them may be checked using some form of protractor or for small deviations, a spirit level.

Roundness: a part is round when all points on its circumference are equidistant from its axis. The simplest check for roundness is to measure directly at a number of diametrically opposite points around the circumference using a micrometer or vernier caliper.

Concentricity: is the relative position of one diameter with another. To be concentric both diameters should lie on the same centreline. The simplest check for concentricity is using a vee block and dial test indicator.

Accuracy of form: it is possible for a number of parts to be manufactured within the

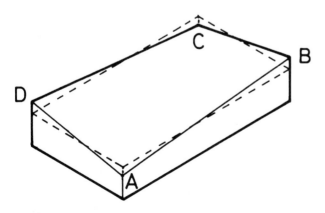

Figure 4.2 Error in flatness of a surface.

prescribed limits of size but not be suitable for their purpose. This is the result of errors in geometry, or form, which can conflict with dimensions of size, e.g. a shaft may be tapered along its length, still be within the limits of size at each end, but not function correctly.

If form has to be controlled to a finer degree than can be achieved by the normal production method and limits of size, then the form may need to be the subject of geometrical tolerancing.

Profile: is the outline of a component and is checked using a template or profile gauge. More precise profiles may need to be checked using an optical projector where an enlarged image of the component is projected on a screen and checked against a profile drawing or transparency of the same magnification.

Relative position: where a number of parts have to function together, their relative position may be extremely important. This is the case with machine tools where the relative position and movement of the main components will determine the subsequent accuracy of the finish machined article. The tests for relative position are carried out using a wide variety of test equipment. Figure 4.3 shows a test to check the parallelism of the spindle of a horizontal milling machine, i.e. the relative position of the spindle to the machine table movement.

Surface roughness: no manufactured surface, however it may appear to the naked eye, is absolutely perfect. The degree of smoothness or roughness of a surface depends on the height and width of a series of peaks and valleys which gives the surface a certain texture. This surface texture is characteristic of the method used to produce it. For example, surfaces produced by cutting tools have tool marks in well-defined directions controlled by the method of cutting, and equally spaced according to the feed rates used.

The control of surface texture is necessary to obtain a surface of known type and roughness value which experience has shown to be most suitable to give long life, fatigue resistance, maximum efficiency, and interchangeability at the lowest cost for a particular application. This is not necessarily achieved by the finest surface. For example, two surfaces required to slide over each other would not function if finished to the same high degree as the surface of a gauge block – they would not slide but simply wring together. At the opposite

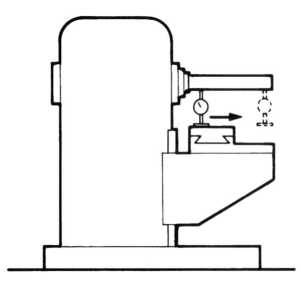

Figure 4.3 Parallelism check.

Figure 4.4 Surface roughness comparison specimens.

Table 4.1

Process	Roughness values (μm R_a)												
	0.0125	0.025	0.05	0.1	0.2	0.4	0.8	1.6	3.2	6.3	12.5	25	50
Superfinishing				■	■								
Lapping			■	■	■	■							
Diamond turning					■	■							
Honing					■	■	■						
Grinding					■	■	■	■					
Turning							■	■	■	■			
Boring							■	■	■	■			
Die-casting								■					
Broaching								■	■				
Reaming								■	■				
Milling								■	■	■			
Investment casting									■				
Drilling									■				
Shaping									■	■	■		
Shell moulding										■			
Sawing										■	■		
Sand casting												■	

extreme, the same two sliding surfaces with a very poor texture would wear quickly. The cost of producing these extremes of surface would also vary greatly.

The method of assessing surface texture is either by sophisticated stylus type equipment or more simply by surface roughness comparison specimens. Comparison specimens (see Fig. 4.4) are used to give engineers, draughtsmen/women, and machine operators an idea of the relation of the feel and appearance of common machined surfaces to their numerical roughness value. By visual examination and by scratching the surface with a finger nail, a comparison can be made between the specimen and the workpiece surface being dealt with.

Some average surface roughness values obtainable from common production processes are shown in Table 4.1.

Length measuring equipment

We know now that some form of precise measurement is necessary and that to achieve any degree of precision requires measuring equipment related to the same standard of length. Having produced the measuring equipment to a high degree of accuracy, it must be used correctly. You must be able to assess the correctness of size of work by choosing the most appropriate instrument and adopt a sensitive touch or 'feel' between the instrument and work being measured. This feel can only be developed from experience of using the equipment, although some instruments do have an aid such as the ratchet stop on some micrometers.

Precision steel rule

These are made from hardened and tempered stainless steel, photoetched for extreme accuracy and have a non-glare satin chrome finish. Rules are available in lengths of 150 mm and 300 mm and graduations may be along each edge of both faces usually in millimetres and half millimetres.

Accuracy of measurement depends on the quality of the rule and the skill of the operator. The width of the lines on a high quality rule are quite fine and accuracies of around 0.15 mm can be achieved but an accuracy of double this can more realistically be achieved.

When measuring between edges, the most common method is to line up the end of the rule on one edge and sight a graduation relative to the other edge. It is more accurate if you butt the rule against a block (see Fig. 4.5a). When measuring a distance between two lines it is easier to work between graduations on the rule (Fig. 4.5b).

Rules are precision measuring instruments and should be treated as such. They should not be used to scrape paint or grease off surfaces, to prise off tin lids nor used as a screwdriver.

Where greater accuracy is required than can be obtained by using a rule then micrometers or vernier instruments have to be used.

Vernier instruments

All instruments employing a vernier consist of two scales: one moving and one fixed. The fixed scale is graduated in millimetres, every 10 divisions equalling 10 mm, and is numbered 0, 1, 2, 3, 4 up to the capacity of the instrument. The moving or vernier scale is divided into 50 equal parts which occupy the same length as 49 divisions or 49 mm on the fixed scale (see Fig. 4.6). This means that the distance between each graduation on the vernier scale is $^{49}/_{50}$ mm = 0.98 mm, or 0.02 mm less than each division on the fixed scale (see Fig. 4.7a).

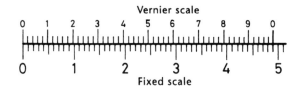

Figure 4.6 Vernier scale.

If the two scales initially have their zeros in line and the vernier scale is then moved so that its first graduation is lined up with a graduation on the fixed scale, the zero on the vernier scale will have moved 0.02 mm (Fig. 4.7b). If the second graduation is lined up, the zero on

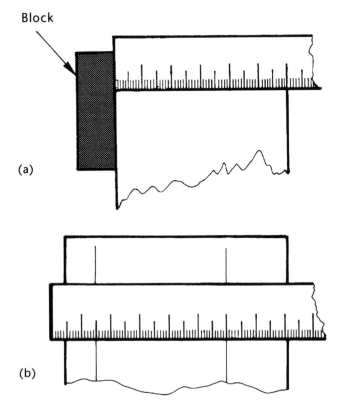

Figure 4.5 Using a rule.

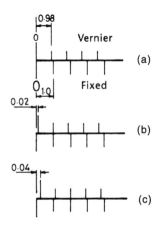

Figure 4.7 Vernier scale readings.

the vernier scale will have moved 0.04 mm (Fig. 4.7c), and so on. If graduation 50 is lined up, the zero will have moved 50 × 0.02 = 1 mm.

Since each division on the vernier scale represents 0.02 mm, five divisions represent 5 × 0.02 = 0.1 mm. Every fifth division on this scale is marked 1 representing 0.1 mm, 2 representing 0.2 mm, and so on (Fig. 4.6).

To take a reading, note how many milli metres the zero on the vernier scale is from zero on the fixed scale. Then note the number of divisions on the vernier scale from zero to a line which exactly coincides with a line on the fixed scale.

In the reading shown in Fig. 4.8(a) the vernier scale has moved 40 mm to the right. The eleventh line coincides with a line on the fixed scale, therefore 11 × 0.02 = 0.22 mm is added to the reading on the fixed scale, giving a total reading of 40.22 mm.

Similarly, in Fig. 4.8(b) the vernier scale has moved 120 mm to the right plus 3 mm and the sixth line coincides, therefore, 6 × 0.02 = 0.12 mm is added to 123 mm, giving a total of 123.12 mm.

It follows that if one part of a measuring instrument is attached to the fixed scale and another part to the moving scale, we have an instrument capable of measuring to 0.02 mm.

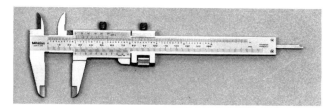

Figure 4.9 Vernier caliper.

ing screws A and B (Fig. 4.11). Move the sliding jaw along the beam until it contacts the surface of the work being measured. Tighten locking screw B. Adjust nut C until the correct 'feel' is obtained, then tighten locking screw A. Re-check 'feel' to ensure that nothing has moved. When you are satisfied, take the reading on the instrument.

Dial calipers (Fig. 4.12) are a form of vernier caliper where readings of 1 mm steps are taken from the vernier beam and sub-divisions of this are read direct on a dial graduated in 0.02 mm divisions

The electronic caliper shown in (Fig. 4.13) is a modern-designed precision measuring instrument which has an electronic measuring

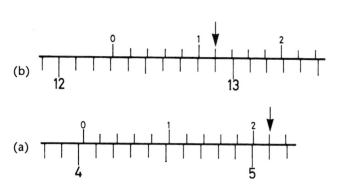

Figure 4.8 Vernier readings.

Vernier caliper: The most common instrument using the above principle is the vernier caliper (see Fig. 4.9). These instruments are capable of external, internal, step and depth measurements (Fig. 4.10) and are available in a range of measuring capacities from 150 mm to 1000 mm.

To take a measurement, slacken both lock-

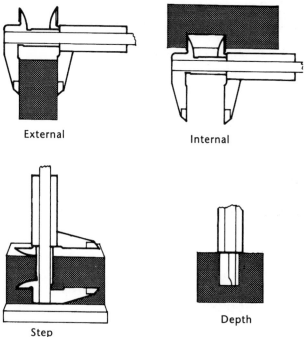

Figure 4.10 External, internal, step and depth measurement.

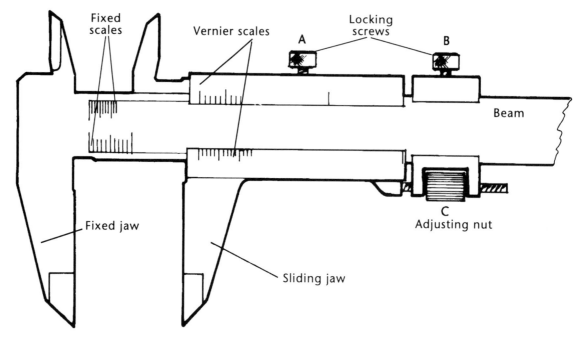

Figure 4.11 Vernier caliper adjustment.

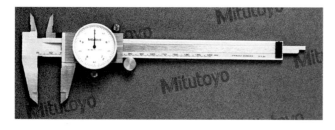

Figure 4.12 Dial caliper.

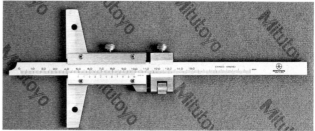

Figure 4.14 Vernier depth gauge.

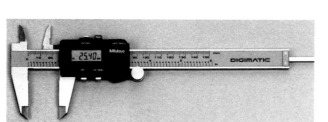

Figure 4.13 Electronic caliper.

moving scale is tee-shaped to provide a substantial base and datum from which readings are taken. The instrument reading is the amount which the rule sticks out beyond the base. Depth gauges are available in a range of capacities from 150 mm to 300 mm.

unit with an LCD digital readout giving direct readings in imperial or metric units with a resolution of 0.0005″ or 0.01 mm.

Vernier depth gauge: accurate depths can be measured using the vernier depth gauge (Fig. 4.14) again employing the same principles. The fixed scale is similar to a narrow rule. The

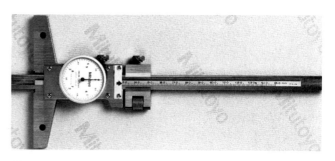

Figure 4.15 Dial depth gauge.

These are also available as an easy-to-read dial depth gauge (Fig. 4.15) and an electronic model with an LCD digital readout (Fig. 4.16) operating in the same way as the caliper models.

Figure 4.16 Electronic depth gauge.

Micrometers

The micrometer relies for its measuring accuracy on the accuracy of the spindle screw thread. The spindle is rotated in a fixed nut by means of the thimble, which opens and closes the distance between the ends of the spindle and anvil (see Fig. 4.17). The pitch of the spindle thread, i.e. the distance between two consecutive thread forms, is 0.5 mm. This means that, for one revolution, the spindle and the thimble attached to it will move a longitudinal distance of 0.5 mm.

On a 0–25 mm micrometer, the sleeve around which the thimble rotates, has a longitudinal line graduated in mm from 0 to 25 mm on one side of the line and subdivided in 0.5 mm intervals on the other side of the line.

The edge of the thimble is graduated in 50 divisions numbered 0, 5, 10, up to 45, then 0. Since one revolution of the thimble advances the spindle 0.5 mm, one graduation on the thimble must equal 0.5 ÷ 50 mm = 0.01 mm. A reading is therefore the number of 1 mm and 0.5 mm divisions on the sleeve uncovered by the thimble plus the hundredths of a millimetre indicated by the line on the thimble coinciding with the longitudinal line on the sleeve.

In the reading shown in Fig. 4.18(a), the thimble has uncovered 9 mm on the sleeve. The thimble graduation lined up with the longitudinal line on the sleeve is 44 = 44 × 0.01 = 0.44. The total reading is therefore 9.44 mm.

Similarly in Fig. 4.18(b) the thimble has uncovered 16 mm and 0.5 mm and the thimble is lined up with graduation 27 = 27 × 0.01 = 0.27 mm, giving a total reading of 16.77 mm.

Greater accuracy can be obtained with external micrometers by providing a vernier scale on the sleeve. The vernier consists of five divisions on the sleeve, numbered 0, 2, 4, 6, 8, 0, these occupying the same space as nine divisions on the thimble, (Fig. 4.19a). Each division on the vernier is therefore equal to 0.09 ÷ 5 = 0.018 mm. This is 0.002 mm less than two divisions on the thimble.

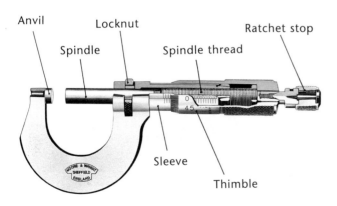

Figure 4.17 External micrometer.

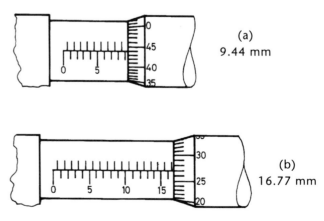

(a)
9.44 mm

(b)
16.77 mm

Figure 4.18 Micrometer readings.

To take a reading from a vernier micro-meter, note the number of 1 mm and 0.5 mm divisions uncovered on the sleeve and the hundredths of a millimetre on the thimble as with an ordinary micrometer. You may find that the graduation on the thimble does not exactly coincide with the longitudinal line on the sleeve, and this difference is obtained from the vernier. Look at the vernier and see which graduation coincides with a graduation on the thimble. If it is the graduation marked 2, then add 0.002 mm to your reading (Fig. 4.19b); if it is the graduation marked 4, then add 0.004 mm (Fig. 4.19c); and so on.

External micrometers with fixed anvils are available with capacities ranging from 0–13 mm to 575–600 mm. External micro-meters with interchangeable anvils, (Fig. 4.20) provide an extended range from two to six times greater than the fixed-anvil types. The smallest capacity is 0–50 mm and the largest 900–1000 mm. To ensure accurate setting of the interchangeable anvils, setting gauges are supplied with each instrument.

Micrometers are available which give a direct mechanical read-out on a counter (Fig.

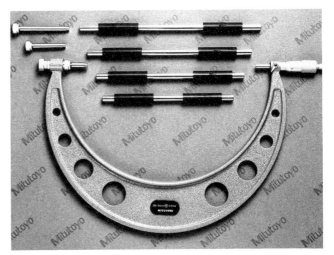

Figure 4.20 External micrometer with interchangeable anvils.

4.21). The smallest model of 0–25 mm has a resolution of 0.001 mm and the largest of 125–150 mm a resolution of 0.01 mm.

Modern-designed external micrometers (Fig. 4.22) are available with an LCD digital readout giving direct readings in either imperial or metric units with a resolution of 0.00005″ or 0.001 mm.

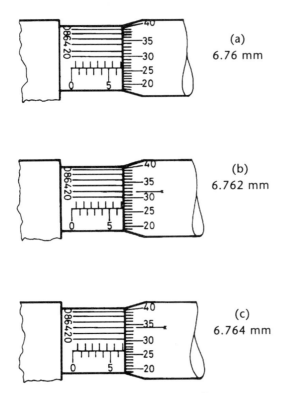

(a) 6.76 mm

(b) 6.762 mm

(c) 6.764 mm

Figure 4.19 Vernier micrometer readings.

Figure 4.21 Direct reading digital micrometer

Figure 4.22 Electronic external micrometer.

Internal micrometer: The internal micrometer, (Fig. 4.23) is designed for inside measurement and consists of a micrometer measuring head to which may be added extension rods to cover a wide range of measurements and a spacing collar to make up for the limited range of the micrometer head. Micrometers of less than 300 mm are supplied with a handle to reach into deep holes. Each extension rod is marked with the respective capacity of the micrometer when that particular rod is used. The smallest size is 25–50 mm with a measuring range of 7 mm. The next size covers 50–200 mm with a measuring range of 13 mm, while the largest covers 200–1000 mm with a measuring range of 25 mm.

Readings are taken in the same way as described for the external micrometer, although, as already stated, the measuring range of the micrometer head is reduced.

Great care must be taken when using this instrument, as each of the measuring anvils has a spherical end, resulting in point contact. Experience in use is essential to develop a 'feel', and the instrument must be moved slightly back and forth and up and down to ensure that the measurement is taken across the widest point.

Depth micrometer: the depth micrometer (Fig. 4.24) is used for measuring the depths of holes, slots, recesses, and similar applications. Two types are available: one with a fixed spindle and a capacity of 0–25 mm, the other with interchangeable rods giving a measuring

capacity up to 300 mm. The interchangeable rods are fitted into the instrument by unscrewing the top part of the thimble and sliding the rod in place, ensuring that the top face of the thimble and the underside of the rod are perfectly clean. The top of the thimble is then replaced and holds the rod in place. Each rod is marked with its respective size.

The micrometer principle is the same as for the other instruments; however, the readings with this instrument increase as the thimble is screwed on, resulting in the numbering of sleeve and thimble graduations in the opposite direction to those on the external and internal micrometers. To take a reading, you must note the 1 mm and 0.5 mm divisions covered by the thimble and add to this the hundredths of a millimetre indicated by the line on the thimble coinciding with the longitudinal line on the sleeve.

In the reading shown in Fig. 4.25(a) the thimble has covered up 13 mm and not quite reached 13.5. The line on the thimble coinciding with the longitudinal line on the sleeve is 44, so $44 \times 0.01 = 0.44$ mm is added, giving a total reading of 13.44 mm. Similarly, in Fig. 4.25(b) the thimble has just covered 17 mm and line 3 on the thimble coincides, giving a total reading of 17.03 mm.

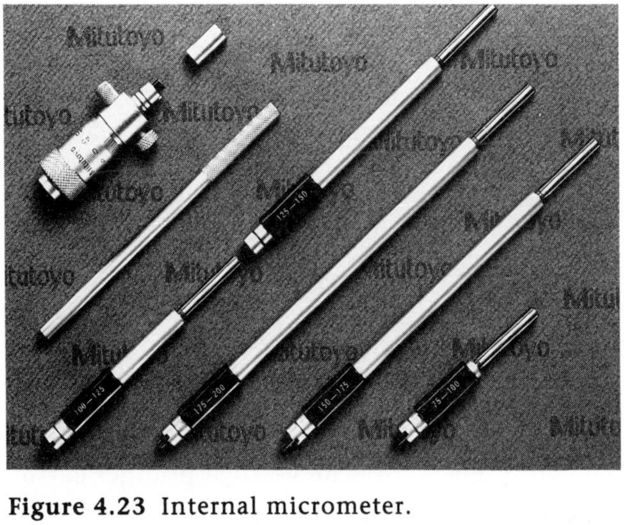

Figure 4.23 Internal micrometer.

Figure 4.24 Depth micrometer.

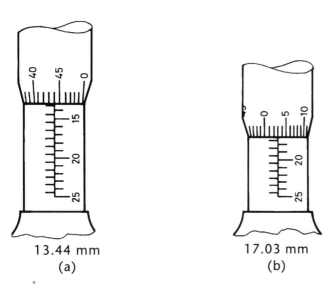

13.44 mm
(a)

17.03 mm
(b)

Figure 4.25 Depth micrometer readings.

Figure 4.27 Electronic depth micrometer.

Depth micrometers are available giving a direct mechanical read-out on a counter with a resolution of 0.01 mm (see Fig. 4.26).

Depth micrometers are also available as an electronic model with an LCD digital readout (Fig. 4.27) giving direct readings in imperial or metric units with a resolution of 0.0001″ or 0.001 mm.

Dial indicators

Dial indicators magnify small movements of a plunger or lever and show this magnified movement by means of a pointer on a graduated dial. This direct reading from the pointer and graduated dial gives the operator a quick, complete, and accurate picture of the condition of the item under test. Dial indicators are used to check the dimensional accuracy of workpieces in conjunction with other equipment such as gauge blocks, to check straightness and alignments of machines and equipment, to set workpieces in machines to ensure parallelism and concentricity and for a host of other uses too numerous to list completely.

The mechanism of a dial indicator is similar to that of a watch and, although made for workshop use, care should be taken to avoid dropping or knocking it in any way. Slight damage to the mechanism can lead to sticking which may result in incorrect or inconsistent readings.

Figure 4.26 Direct reading digital depth micrometer.

Plunger-type instruments

The most common instrument of this type is shown in Fig. 4.28. The vertical plunger carries a rack which operates a system of gears for

Figure 4.28 Plunger type dial indicator.

Figure 4.29 Dial test indicator and stand.

magnification to the pointer. The dial is attached to the outer rim, known as the bezel, and can be rotated so that zero can be set irrespective of the initial pointer position. A clamp is also supplied to prevent the bezel moving once it has been set to zero. The dial divisions are usually 0.01 or 0.002 mm, with an operating range between 8 and 20 mm, although instruments with greater ranges are available.

In conjunction with a robust stand or surface gauge (Fig. 4.29) this instrument can be used to check straightness, concentricity, as well as workpiece heights and roundness.

It may not always be possible to have the dial of this type facing the operator, which may create problems in reading the instrument or safety problems if the operator has to bend over equipment or a machine. An instrument which can be used to overcome these difficulties is the back plunger type shown in Fig. 4.30. The reading can be easily seen by viewing above the instrument. The direction of the plunger movement restricts the range to about 3 mm.

Modern-design electronic plunger type instruments (Fig. 4.31) are available with an LCD digital readout giving direct readings in imperial or metric units with a resolution of 0.0005″ or 0.01 mm. These instruments can be zeroed at any point within the range.

Lever-type instruments

The lever type of instrument is shown in Fig. 4.32. Due to the leverage system, the range of this type is not as great as that of the plunger type and is usually 0.5 or 0.8 mm. The dial divisions are 0.01 or 0.005 mm, and again the dial is adjustable to set zero. The greatest advantage of this type is the small space within which it can work. Another added

Figure 4.30 Back plunger type dial indicator.

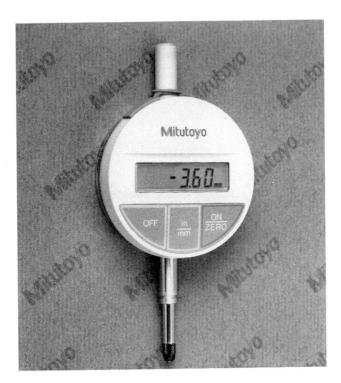

Figure 4.31 Electronic type indicator.

Figure 4.32 Lever type dial indicator.

advantage is an automatic reversal system which results in movement above or below the contact stylus registering on the dial pointer. This facility, together with the contact stylus being able to swing at an angle, means that checks can be made under a step as well as on top (Fig. 4.33).

Angle measuring equipment

Two surfaces are square when they are at right angles to each other. Thus the determination of squareness is one of angular measurement. The workshop standard for checking right angles is the square of which there are a number of types including engineer's squares and cylindrical squares.

Engineer's squares (Fig. 4.34) consist of a stock and a blade and are designated by a size which is the length from the tip of the blade to the inner working face of the stock. Available sizes range from 75 to 1050 mm. Three grades of accuracy are specified – AA, A, and B – with grade AA the most accurate. Engineer's squares are made of good quality steel, with the working faces of grades AA and A hardened and stabilised. All working surfaces of the blade and the stock are lapped, finely ground, or polished to the accuracy specified for each grade.

To check the squareness of two surfaces of a workpiece using an engineer's square, the stock is placed on one face and the edge of the

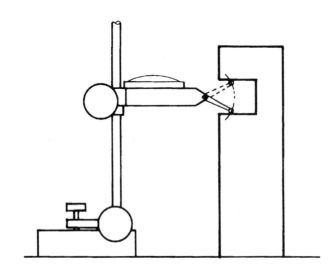

Figure 4.33 Lever type dial indicator application.

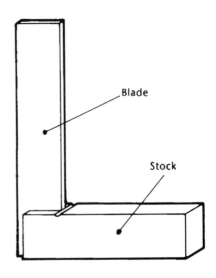

Figure 4.34 Engineer's square.

blade is rested on the other. Any error in squareness can be seen by the amount of light between the surface and the underside of the blade. This type of check only tells you whether or not the surfaces are square to each other, however, and it is difficult to judge the magnitude of any error.

When accurate results are required, the workpiece and square may be placed on a surface plate and the square slid gently into contact with the surface to be checked. The point of contact can be viewed against a well illuminated background. If a tapering slit of light can be seen, the magnitude of the error present can be checked using gauge blocks or feeler gauges at the top and bottom of the surface (Fig. 4.35), the difference between the two gauge blocks or feeler gauges being the total error in squareness.

Cylindrical squares (Fig. 4.36), of circular section, are designated by their length and are available in lengths from 75 mm up to 750 mm. One grade of accuracy is specified – grade AA. Cylindrical squares are made from high-quality steel hardened and stabilised, close-grained plain or alloy cast iron, or granite. All external surfaces are finished by lapping or fine grinding.

Cylindrical squares, because of their high accuracy, are used for checking the squareness of engineers squares and for precise work with flat surfaces, since line contact by the cylindrical surface gives greater sensitivity.

Angles other than right angles have to be checked using some form of protractor. There is no absolute standard for angular measurement in the same way as there is for linear measurement since the requirement is simply to divide

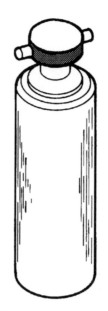

Figure 4.36 Cylindrical square.

a circle into a number of equal parts. Thus a circle divided into 360 equal parts gives a degree – a degree further divided into 60 equal parts gives a minute – and a minute further divided into 60 equal parts gives a second. Anything finer than this is given as fractions of a second.

Protractor

The simplest form of protractor is shown in Fig. 4.37 where the rule pivots around a body which is graduated 0–180° in each direction in divisions of one degree. The body is placed against

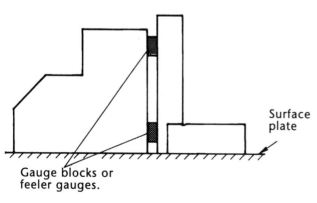

Figure 4.35 Checking squareness.

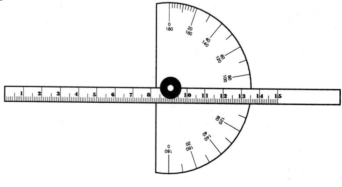

Figure 4.37 Simple protractor.

the datum surface and the blade rotated to coincide with the face being measured. The angle reading is taken directly from the scale. Although greater accuracy can be achieved by an experienced person, an accuracy of 30 minutes (½ degree) is realistic. The protractor head of a combination set can be used in the same way see Fig. 5.14 on page 71. Where greater accuracy is required, angles can be measured using a vernier bevel protractor.

Vernier bevel protractor

As well as linear measurement, vernier scales can equally well be used to determine angular measurement. The vernier bevel protractor (Fig. 4.38) again uses the principle of two scales, one moving and one fixed. The fixed scale is graduated in degrees, every 10 degrees being numbered 0, 10, 20, 30, etc. The moving or vernier scale is divided into 12 equal parts which occupy the same space as 23 degrees on the fixed scale (Fig. 4.39). This means that each division on the vernier scale is $^{23}/_{12}$ degrees = $1^{11}/_{12}$ degrees or 1 degree 55 minutes. This is 5 minutes less than two divisions on the fixed scale (Fig. 4.40a).

If the two scales initially have their zeros in line and the vernier scale is then moved so that its first graduation lines up with the 2 degree graduation on the fixed scale, the zero

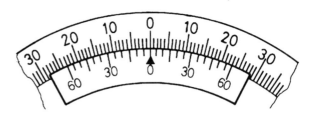

Figure 4.39 Vernier protractor scale.

on the vernier scale will have moved 5 minutes (Fig. 4.40b). Likewise, the second graduation of the vernier lined up with the 4 degree graduation will result in the vernier scale zero moving 10 minutes (Fig. 4.40c) and so on until when the twelfth graduation lines up the zero will have moved 12 × 5 = 60 minutes = 1 degree. Since each division on the vernier scale represents 5 minutes, the sixth graduation is

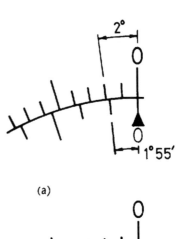

(a)

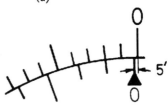

(b)

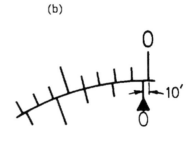

(c)

Figure 4.40 Vernier protractor scale readings.

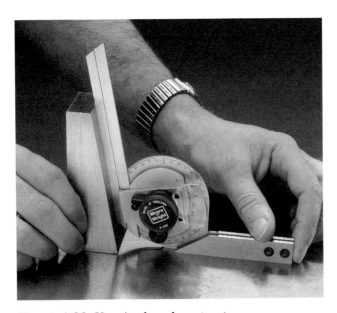

Figure 4.38 Vernier bevel protractor.

numbered to represent 30 minutes and the twelfth to represent 60 minutes.

The stock of the vernier protractor carries the fixed scale. The removable blade is attached to the moving or vernier scale, which has a central screw to lock the scale at any desired position and give angular measurement to an accuracy of 5 minutes.

Since the vernier scale can be rotated in both directions, the fixed scale is graduated from 0–90, 90–0, 0–90, 90–0 through 360°. This requires a vernier scale for each, and therefore the vernier scale is also numbered 0–60 in each direction.

To take a reading, note how many degrees the zero on the vernier scale is from the zero on the fixed scale. Then counting in the same direction, note the number of divisions on the vernier scale from zero to a line which exactly coincides with a line on the fixed scale.

In the reading shown in Fig. 4.41(a) the vernier scale has moved to the left 45 degrees. Counting along the vernier scale in the same direction, i.e. to the left, the seventh line coincides with a line on the fixed scale. Thus 7 × 5 = 35 minutes is added, to give a total reading of 45 degrees 35 minutes.

In the reading shown in Fig. 4.41(b), the vernier scale has moved to the right 28 degrees. Again counting along the vernier scale in the same direction, i.e. to the right, the eleventh line coincides with a line on the fixed scale, giving a total reading of 28 degrees 55 minutes.

Table 4.2 on page 60 shows a summary of measuring instruments.

Terminology of measurement

Indicated size: the actual size indicated by the measuring scales. It does not make any allowance for errors due to wear in the instrument, incorrect application, or use of unnecessary force.

Reading: this is the visual interpretation of the indicated size. It relies on a persons ability to recognise and line up graduations on the instrument. Some instruments are more difficult to read than others. The vernier bevel protractor shown in Fig. 4.38 has a built-in magnifying glass for easier reading.

Reading value: or resolution is the smallest size increment which can be read on a measuring instrument, e.g. a micrometer has a resolution of 0.01 or 0.002 mm with a vernier scale while a vernier caliper has a resolution of 0.02 mm.

Measuring range: this is the difference between the maximum and minimum measuring capacity of an instrument, e.g. a 50–75 mm external micrometer has a measuring range of 25 mm.

Measuring accuracy: this is the expected accuracy of the instrument taking into consideration all the above factors and will always be less than the true capability of the instrument.

Factors affecting accuracy of measurement

Temperature

As stated at the beginning of the chapter, accurate measurement and calibration would be carried out under exacting conditions at 20°C.

(a)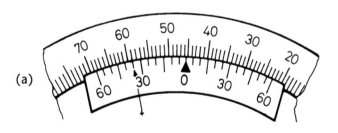

(b)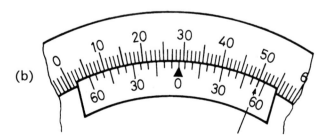

Figure 4.41 Vernier protractor readings.

Table 4.2

Instrument	Advantages	Limitations
Vernier calipers	Large measuring range on one instrument. Capable of internal, external, step and depth measurements. Resolution of LCD model 0.01 mm	Resolution 0.02 mm. Point of measuring contact not in line with adjusting nut. Jaws can spring. Lack of 'feel'. Length of jaws limits measurement to short distance from end of work piece. No adjustment for wear
Vernier height gauge	Large range on one instrument. Resolution of LCD model 0.01 mm	Resolution 0.02 mm. No adjustment for wear
Vernier depth gauge	Large range on one instrument. Resolution of LCD model 0.01 mm	Resolution 0.02 mm. Lack of 'feel'. No adjustment for wear
Bevel protractor	Accuracy 5 minutes over range of 360°. Will measure internal and external angles	Can be difficult to read the small scales except with the aid of a magnifying lens
External micrometer	Resolution 0.01 mm or, with vernier 0.002 mm or LCD model 0.001 mm. Adjustable for wear. Ratchet or friction thimble available to aid constant 'feel'	Micrometer head limited to 25 mm range. Separate instruments required in steps of 25 mm or by using interchangeable anvils
Internal micrometer	Resolution 0.01 mm. Adjustable for wear. Can be used at various points along length of bore	Micrometer head on small sizes limited to 7 and 13 mm range. Extension rods and spacing collar required to extend capacity. Difficult to obtain 'feel'
Depth micrometer	Resolution 0.01 mm or with LCD model 0.001 mm. Adjustable for wear. Ratchet or friction thimble available to aid constant 'feel'	Micrometer head limited to 25 mm range. Interchangeable rods required to extend capacity
Dial indicator	Resolution as high as 0.001 mm. Measuring range up to 80 mm with plunger types. Mechanism ensures constant 'feel'. Easy to read. Quick in use if only comparison is required	Has to be used with gauge blocks to determine measurement. Easily damaged if mishandled. Must be rigidly supported in use

Accuracy of equipment

The accuracy of the measuring instrument used must be of a higher order than the dimension to be measured.

Measurement errors

To avoid measurement errors:

■ Choose measuring instruments most appropriate to the task e.g. vernier calipers are only capable of measuring at the mouth of a bore (Fig. 4.42).

■ Avoid parallax by getting your eye in line with graduation marks and the point being measured e.g. if when using a rule the eye is in line with the point of measurement at A in Fig. 4.43 an accurate reading is obtained – if the eye is at B then the point of measurement coincides with a different rule graduation resulting in an error reading.

■ It is more accurate to achieve a measurement which uses a positive datum. A length may be the distance between two lines (line measurement) or the distance between two faces (end measurement). Examples of this is the use of a rule for line measurement

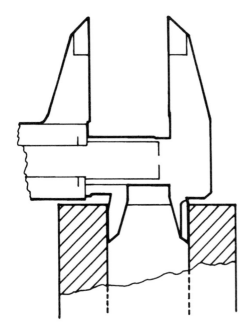

Figure 4.42 Vernier measuring at mouth of bore.

and a micrometer for end measurement. Recognise the difficulty of transferring line measurement to end measurement, e.g. it is more accurate to set calipers across the faces of a block of the correct thickness than to set it to a rule.

- The development of 'feel' is most important. Errors can be brought about by using too much force when taking a measurement. e.g. too much force when using a micrometer will give a error of measurement as well as possible damage to the instrument. Use of too much force when measuring fragile workpieces can distort the part and lead to errors.
- Correct positioning of the measuring instrument is essential, e.g. when using an internal micrometer, the instrument has to be moved slightly back and forth and up and down to ensure an accurate measurement across the widest point.

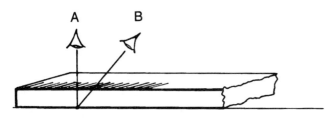

Figure 4.43 Parallax using a rule.

- Cleanliness of workpiece and measuring equipment is essential. It is no good trying to measure a part which is covered in grease, dust or swarf particles.
- Measuring instruments have to be checked and calibrated at regular intervals, e.g. it is no good assuming that because you are using a micrometer it is bound to be correct. Wear does take place through continual use (or misuse) and adjustments are possible on some instruments.
- Measuring instruments should be treated with care at all times. Do not throw them down, leave lying on the bench or machine where they can get damaged nor place other items on top of them.

Dimensional deviation

Engineering workpieces cannot be consistently produced to an exact size. This is due to a number of reasons such as wear on cutting tools, errors in setting up, operator faults, temperature differences, or variations in machine performance. Whatever the reason, allowance must be made for some error. The amount of error which can be tolerated – known as the tolerance – depends on the manufacturing method and on the functional requirements of the workpiece. For example, a workpiece finished by grinding can be consistently made to finer tolerances than one produced on a centre lathe. In a similar way, a workpiece required for agricultural equipment would not require the same fine tolerance required for a wristwatch part. In fact it would be expensive and pointless to produce parts to a greater accuracy than was necessary for the part to function.

Establishing a tolerance for a dimension has the effect of creating two extremes of size, or limits – a maximum limit of size and a minimum limit of size within which the dimension must be maintained.

British Standard BS4500 provides a comprehensive standardised system of limits and fits for engineering purposes. This British Standard relates to tolerances and limits of size for workpieces, and to the fit obtained when two workpieces are to be assembled.

BS4500 is based on a series of tolerances graded to suit all classes of work covering a range of workpiece sizes up to 3150 mm. A series of qualities of tolerance – called tolerance grades – is provided, covering fine tolerances at one end to coarse tolerances at the other. BS4500 provides eighteen tolerance grades, designated IT01, IT0, IT1, IT2, IT3, IT4, IT5, IT6, IT7, IT8, IT9, IT10, IT11, IT12, IT13, IT14, IT15 and IT16. (IT stands for ISO series of tolerances.) The numerical values of these standard tolerances for nominal work sizes up to 500 mm are shown in Table 4.3. You can see from the table, that for a given tolerance grade the magnitude of the tolerance is related to the nominal size, e.g. for tolerance grade IT6 the tolerance for a nominal size of 3 mm is 0.006 mm while the tolerance for a 500 mm nominal size is 0.04 mm. This reflects both the practicalities of manufacture and of measuring.

Typical tolerance grades obtainable from various manufacturing processes are shown in Table 4.4.

Table 4.4. Typical tolerance grades

Tolerance grade	Manufacturing process
IT 01	
0	
1	Lapping
2	
3	Honing and superfinishing
4	
5	Diamond turning and fine grinding
6	Grinding
7	High-quality turning and broaching
8	Turning, boring and reaming
9	High-quality milling
10	Milling
11	Drilling, rough turning and boring, diecasting
12	Shaping and investment casting
13	Drawing and presswork
14	Shell moulding
15	Sand casting
16	Flame cutting

Table 4.3. Standard tolerances (tolerance unit 0.001 mm)

Nominal sizes Over (mm)	To (mm)	IT 01	IT 0	IT 1	IT 2	IT 3	IT 4	IT 5	IT 6	IT 7	IT 8	IT 9	IT 10	IT 11	IT 12	IT 13	IT 14*	IT 15*	IT 16*
–	3	0.3	0.5	0.8	1.2	2	3	4	6	10	14	25	40	60	100	140	250	400	600
3	6	0.4	0.6	1	1.5	2.5	4	5	8	12	18	30	48	75	120	180	300	480	750
6	10	0.4	0.6	1	1.5	2.5	4	6	9	15	22	36	58	90	150	220	360	580	900
10	18	0.5	0.8	1.2	2	3	5	8	11	18	27	43	70	110	180	270	430	700	1100
18	30	0.6	1	1.5	2.5	4	6	9	13	21	33	52	84	130	210	330	520	840	1300
30	50	0.6	1	1.5	2.5	4	7	11	16	25	39	62	100	160	250	390	620	1000	1600
50	80	0.8	1.2	2	3	5	8	13	19	30	46	74	120	190	300	460	740	1200	1900
80	120	1	1.5	2.5	4	6	10	15	22	35	54	87	140	220	350	540	870	1400	2200
120	180	1.2	2	3.5	5	8	12	18	25	40	63	100	160	250	400	630	1000	1600	2500
180	250	2	3	4.5	7	10	14	20	29	46	72	115	185	290	460	720	1150	1850	2900
250	315	2.5	4	6	8	12	16	23	32	52	81	130	210	320	520	810	1300	2100	3200
315	400	3	5	7	9	13	18	25	36	57	89	140	230	360	570	890	1400	2300	3600
400	500	4	6	8	10	15	20	27	40	63	97	155	250	400	630	970	1550	2500	4000

*Not applicable to sizes below 1 mm.

Terminology

Limits of size – the maximum and minimum sizes permitted for a feature.

Maximum limit of size – the greater of the two limits of size.

Minimum limit of size – the smaller of the two limits of size.

Basic size – the size to which the two limits of size are fixed. The basic size is the same for both members of a fit and can be referred to as nominal size.

Upper deviation – the algebraic difference between the maximum limit of size and the corresponding basic size.

Lower deviation – the algebraic difference between the minimum limit of size and the corresponding basic size.

Tolerance – the difference between the maximum limit of size and the minimum limit of size (or, in other words, the algebraic difference between the upper deviation and lower deviation).

Mean size – the dimension which lies mid-way between the maximum limit of size and the minimum limit of size.

Cumulative errors

When a number of workpieces are assembled, the possible overall error is an accumulation of the error of each individual workpiece. For example, imagine five blocks each with an individual tolerance of 0.2 mm placed one on top of the other. The possible overall error or cumulative error could be 1 mm. This probability must be taken into account by the designer at the design stage.

As well as being present in an assembly, cumulative errors may also occur on a single workpiece. In this case the amount of cumulative error depends not only on the tolerance but also on the system of dimensioning. In Fig. 4.44(a) the overall length is subject to the accumulation of errors occurring at each of the steps, i.e. up to 0.6 mm. Whereas in Fig. 4.44(b), the overall length is controlled by a single tolerance of 0.15 mm.

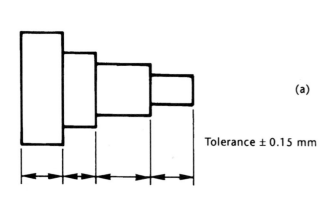

(a)

Tolerance ± 0.15 mm

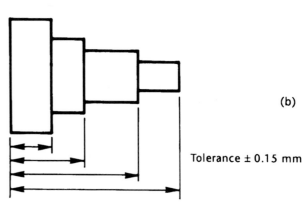

(b)

Tolerance ± 0.15 mm

Figure 4.44 Cumulative error.

Equipment used to aid measuring

Surface table and surface plate

In order to establish a datum from which all measurements are made a reference surface is required. This reference surface takes the form of a large flat surface called a surface table (Fig. 4.45) upon which the measuring equipment is used.

Surface plates are smaller reference surfaces and are placed on a bench for use with smaller workpieces. For general use, both surface tables and surface plates are made from cast iron machined to various grades of accuracy. For high-accuracy inspection work and for use in standards rooms, surface tables and plates made from granite are available.

Vee blocks

Holding circular work during measuring can be simplified by using a vee block (Fig. 4.46). The larger sizes are made from cast iron, the smaller sizes from steel, hardened and ground, and provided with a clamp. They are supplied in pairs marked for identification. The faces are machined to a high degree of accuracy of flatness, squareness, and parallelism, and the 90° vee is central with respect to the side faces and parallel to the base and side faces.

Figure 4.46 Vee blocks.

Figure 4.45 Surface table.

Straight edge

The workshop standard against which the straightness of a line on a surface is compared is the straight edge. Three types of straight edge are available: toolmakers' straight edges, cast iron straight edges, and steel or granite straight edges of rectangular section.

Toolmakers' straight edges are of short length up to 300 mm and are intended for very accurate work. They are made from high quality steel, hardened and ground and have the working edge ground and lapped to a 'knife edge' as shown in Fig. 4.47 section A–A. This type of straight edge is used by placing the knife edge on the work and viewing against a well-illuminated background. If the work is perfectly straight at that position, then no white light should be visible at any point along the length of the straight edge. It is claimed that this type of test is sensitive to within 1 μm.

Rectangular section straight edges, with or without a bevel edge, are available in lengths from 300 to 1800 mm and are used for larger work in the same way as the toolmakers' straight edge.

Cast iron straight edges of bow shape (Fig. 4.48a), and I-section (Fig. 4.48b) are available. Two grades of accuracy are provided for each type – grade A and grade B – with grade A the more accurate. These types of straight edge are used extensively to check the straightness of machine tool slides and slideways. This is done by smearing a thin even layer of engineer's blue on the working surface of the straight edge, placing the straight edge on the surface to be checked and sliding it backwards and forwards a few times. Engineer's blue from

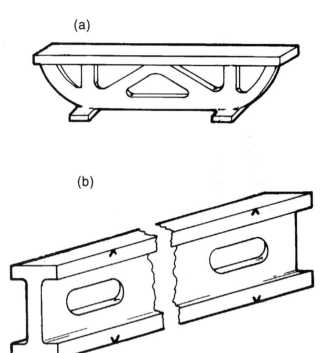

(a)

(b)

Figure 4.48 Cast iron straight edges.

the straight edge is transferred to the work surface, giving an indication of straightness by the amount of blue present. Due to the width of the working face of the straight edge, an indication of flatness is also given. When a straight edge is used on edge, it is likely to deflect under its own weight. For minimum deflection, a straight edge must be supported at two points located two-ninths of its length from each end. For this reason, rectangular and I-section straight edges have arrows together with the word 'support' engraved on their side faces to indicate the points at which the straight edge should be supported for minimum deflection under its own weight.

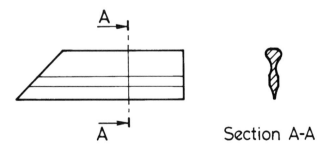

Figure 4.47 Toolmaker's straight edge.

Feeler gauges

Feeler gauges (or thickness gauges), are available as sets of blades (or leaves) made from hardened and tempered steel and of accurate thickness. A typical set (shown in Fig. 4.49) covers a range of 20 thicknesses from 0.05 to 1 mm in steps of 0.05 mm. A feeler gauge is used to check the size of small gaps or error between two objects, i.e. between the workpiece and a reference standard, for example, under a straight edge for checking error in straightness or in conjunction with a square for checking squareness. The method used to check the gap is by trial and error starting with a thinner gauge and working up until the correct thickness blade will just slide with slight resistance between the gap. Care must be taken to avoid bending or wrinkling the thinner blades.

Spirit levels

A spirit level is used to measure an angle relative to the horizontal, e.g. setting up work on a surface table parallel to the table surface. Great care has to be taken using a spirit level in this way since the reference surface must itself be level. If the reference surface is not level then any work set level upon it will not be parallel to the reference surface.

Precision engineers' levels (Fig. 4.50) are extremely sensitive and accurate and are widely used to carry out geometrical tests on machine tools and to determine straightness and flatness of surface tables and plates. The model shown has a length of 200 mm and each division represents a deviation of 0.1 mm over a length of 1 metre.

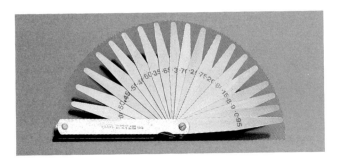

Figure 4.49 Feeler gauges.

Figure 4.50 Precison engineer's level.

5

Marking out

Marking out is the scratching of lines on the surface of a workpiece, known as scribing, and is usually carried out only on a single workpiece or a small number of workpieces. The two main purposes of marking out are:

a) to indicate the workpiece outline or the position of holes, slots, etc. If the excess material will have to be removed, a guide is given for the extent to which hacksawing or filing can be carried out;

b) to provide a guide to setting up the workpiece on a machine. The workpiece is set up relative to the marking out and is then machined. This is especially important when a datum has to be established when castings and forgings are to be machined.

It is important to note that the scribed lines are only a guide, and any accurate dimension must be finally checked by measuring.

Datum

The function of a datum is to establish a reference position from which all dimensions are taken and hence all measurements are made. The datum may be a point, an edge or a centre line, depending on the shape of the workpiece. For any plane surface, two datums are required to position a point and these are usually at right angles to each other.

Figure 5.1 shows a workpiece where the datum is a point; Fig. 5.2 shows a workpiece where both datums are edges; Fig. 5.3 shows a workpiece where both datums are centre lines;

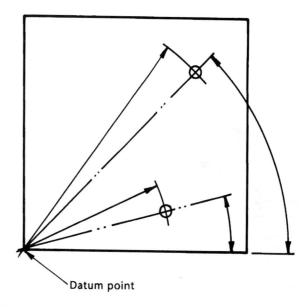

Figure 5.1 Datum point.

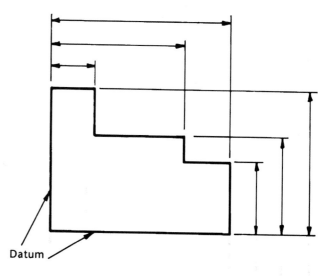

Figure 5.2 Datum edges.

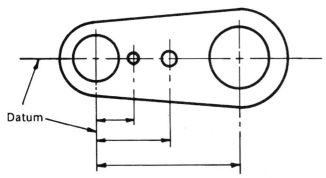

Figure 5.3 Datum centre lines.

and, Fig. 5.4 shows a workpiece where one datum is an edge and the other is a centre line.

The datums are established by the draughtsman when the drawing is being dimensioned and, since marking out is merely transferring drawing dimensions to the workpiece, the same datums are used.

Co-ordinates

The draughtsman can dimension drawings in one of two ways:

■ **Rectangular co-ordinates:** where the dimensions are taken relative to the datums at right angles to each other, i.e. the general pattern is rectangular. This is the method shown in Figs 5.2 and 5.4.

■ **Polar co-ordinates:** where the dimension is measured along a radial line from the

datum. This is shown in Fig. 5.1 Marking out polar co-ordinates requires not only accuracy of the dimension along the radial line but accuracy of the angle itself. As the polar distance increases, any slight angular error will effectively increase the inaccuracy of the final position.

The possibility of error is less with rectangular co-ordinates, and the polar co-ordinate dimensions shown in Fig. 5.1 could be redrawn as rectangular co-ordinates as shown in Fig. 5.5.

Marking out equipment
Surface table and surface plate

Used as a reference working surface to provide a datum upon which all the marking out equipment is used (see also page 64 and Fig. 4.45).

Parallels (Fig. 5.6)

The workpiece can be set on parallels to raise it off the reference surface and still maintain parallelism. Parallels are made in pairs to precisely the same dimensions, from hardened steel, finish ground, with their opposite faces parallel and adjacent faces square. A variety of sizes should be available for use when marking out.

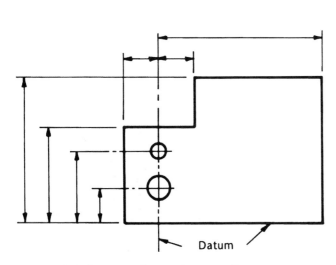

Figure 5.4 Datum edge and centre line.

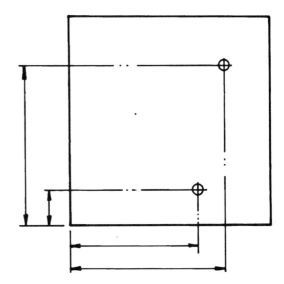

Figure 5.5 Rectangular co-ordinates.

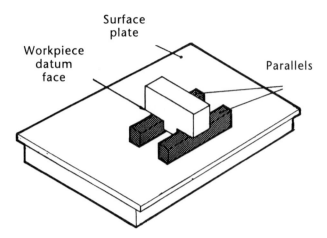

Figure 5.6 Surface plate and parallels.

Jacks and wedges

When a forging or casting, has to be marked out, which has an uneven surface or is awkward in shape, it is still essential to maintain the datum relative to the reference surface. Uneven surfaces can be prevented from rocking and kept on a parallel plane by slipping in thin steel or wooden wedges (Fig. 5.7) at appropriate positions. Awkward shapes can be kept in the correct position by support from adjustable jacks (see Fig. 5.8).

Figure 5.7 Wedge.

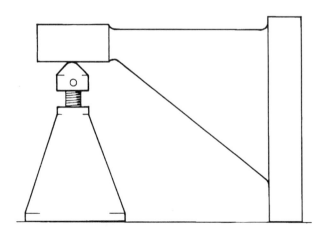

Figure 5.8 Jack used to support.

Angle plate

When the workpiece has to be positioned at 90° to the reference surface, it can be clamped to an angle plate (Fig. 5.9). Angle plates are usually made from cast iron and the edges and faces are accurately machined flat, square and parallel. Slots are provided in the faces for easy clamping of the workpiece. Angle plates may be plain or adjustable.

Figure 5.9 Angle plate and surface gauge.

Vee block

Used to hold circular workpieces when marking out (see Fig. 5.10 and also see page 64 and Fig. 4.46).

Figure 5.10 Vee block in use.

Engineer's square *(Fig. 5.11)*

An engineer's square is used when setting the workpiece square to the reference surface, (see Fig. 5.12) or when scribing lines square to the datum edge (Fig. 5.13). The square consists of a stock and blade made from hardened steel and ground on all faces and edges to give a high degree of accuracy in straightness, parallelism, and squareness. It is available in a variety of blade lengths.

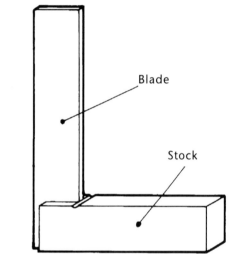

Blade

Stock

Figure 5.11 Engineer's square.

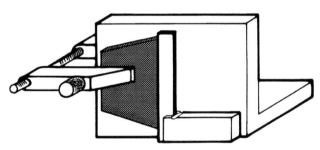

Figure 5.12 Setting workpiece square to reference surface.

Combination set *(Fig. 5.14)*

The combination set consists of a graduated hardened steel rule on which any of three separate heads – protractor, square, or centre head – can be mounted. The rule has a slot in which each head slides and can be locked at any position along its length.

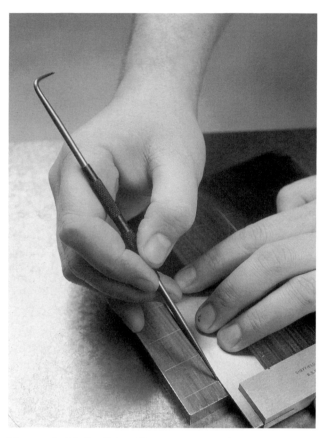

Figure 5.13 Scribing line square to datum.

Protractor head (Fig. 5.15): this head is graduated from 0 to 180°, is adjustable through this range, and is used when scribing lines at an angle to a workpiece datum.

Square head (Fig. 5.16a): this head is used in the same way as an engineer's square but, because the rule is adjustable, it is not as accurate. A second face is provided at 45° (Fig. 5.16b). A spirit level is incorporated which is useful when setting workpieces such as castings level with the reference surface. Turned on end, this head can also be used as a depth gauge (see Fig. 5.16c).

Centre head (Fig. 5.17): with this head the blade passes through the centre of the vee and is used to mark out the centre of a circular workpiece or round bar.

Marking dye

On surfaces of metal other than bright metals, scribed lines may not be clearly visible. In such cases the surface can be brushed or

Figure 5.14 Combination set.

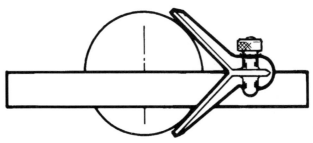

Figure 5.17 Centre head.

sprayed with a quick drying coloured dye before marking out. This provides a good contrast, making the scribed lines easy to see.

Scriber (Fig. 5.18)

The scriber is used to scribe all lines on a metal surface and is made from hardened and tempered steel, ground to a fine point which should always be kept sharp to give well-defined lines

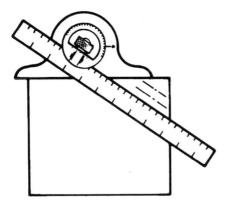

Figure 5.15 Protractor head.

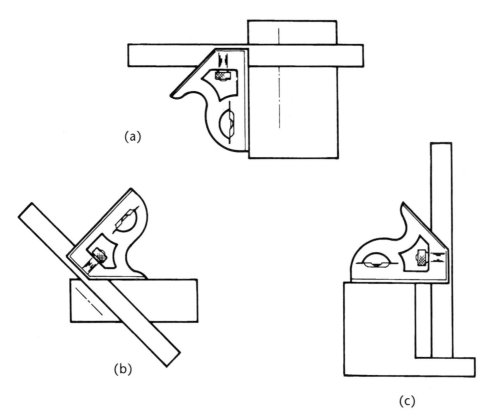

(a)

(b)

(c)

Figure 5.16 Square head.

Figure 5.18 Scriber.

Precision steel rule

Always use a good quality steel rule. With careful use accuracies of around 0.15 mm can be achieved although an accuracy of double this can more realistically be expected (also see page 48).

Surface gauge (*Fig. 5.9*)

The surface gauge, also known as a scribing block, is used in conjunction with a scriber to mark out lines on the workpiece parallel with the reference surface. The height of the scriber is adjustable and is set in conjunction with a steel rule. The expected accuracy from this set up will be around 0.3 mm but with care this can be improved.

Vernier height gauge (*Fig. 5.19*)

Where greater accuracy is required than can be achieved using a surface gauge, marking out can be done using a vernier height gauge. The vernier scale carries a jaw upon which various attachments can be clamped. When marking out, a chisel pointed scribing blade is fitted. Care should be taken to allow for the thickness of the jaw, depending on whether the scribing blade is clamped on top or under the jaw. The precise thickness of the jaw is marked on each instrument. These instruments can be read to an accuracy of 0.02 mm (see page 48) and are available in a range of capacities reading from 0 to 1000 mm.

Dividers and trammels

Dividers are used to scribe circles or arcs and to mark off a series of lengths such as hole centres. They are of spring bow construction, each of the two pointed steel legs being hardened and ground to a fine point and capable of scribing a maximum circle of around 150 mm

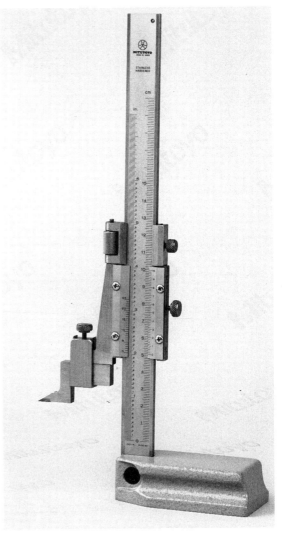

Figure 5.19 Vernier height gauge.

diameter (Fig. 5.20). Larger circles can be scribed using trammels, where the scribing points are adjustable along the length of a beam (Fig. 5.21).

Dividers and trammels are both set in conjunction with a steel rule by placing one point in a convenient graduation line and adjusting the other to coincide with the graduation line the correct distance away.

Hermaphrodite calipers (*Fig. 5.22*)

These combine a straight pointed divider leg with a curved caliper or stepped leg and are used to scribe a line parallel to the edge of a workpiece. They are more commonly known as 'odd-legs' or 'jennies'.

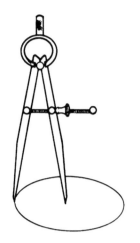

Figure 5.20 Dividers.

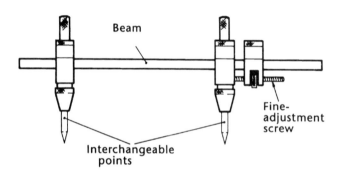

Figure 5.21 Trammels.

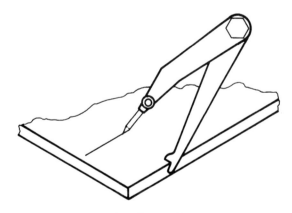

Figure 5.22 Hermaphrodite calipers.

Centre punch *(Fig. 5.23)*

The centre punch is used to provide a centre location for dividers and trammels when scribing circles or arcs, or to show permanently the position of a scribed line by a row of centre dots. The centre dot is also used as a start for small diameter drills.

Centre punches are made from high carbon steel, hardened and tempered with the point ground at 30° when used to provide a centre location for dividers and at 90° for other purposes.

Care should be taken in the use of centre dots on surfaces which are to remain after machining, since, depending upon the depth, they may prove difficult to remove.

Figure 5.23 Centre punch.

Clamps

Clamps are used when the workpiece has to be securely fixed to another piece of equipment, e.g. to the face of an angle plate (Fig. 5.9).

The type most used are toolmaker's clamps (Fig. 5.24), which are adjustable within a range of about 100 mm but will only clamp parallel surfaces. Greater thicknesses can be clamped using 'G' clamps, so named because of their shape (Fig. 5.25). Due to the swivel pad on the end of the clamping screw, the 'G' clamp is also capable of clamping surfaces which are not parallel.

Care should be taken to avoid damage to the surfaces by the clamp.

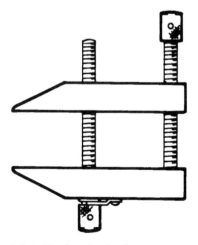

Figure 5.24 Toolmaker's clamp.

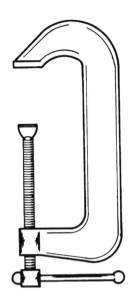

Figure 5.25 'G' clamp.

Examples of marking out

We will now see how to mark out a number of components.

Example 1: component shown in Fig. 5.26

The plate shown at step 1 has been filed to length and width with the edges square and requires the position of the steps to be marked out

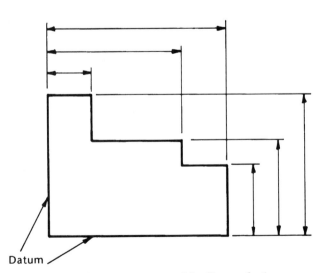

Datum

Figure 5.26 Component used in Example 1.

Step 1 Use a square on one datum edge and measure the distance from the other datum edge using a precision steel rule. Scribe lines.

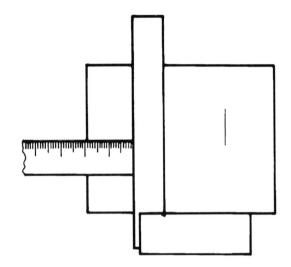

Step 2 Repeat with the square on the second datum edge and scribe lines to intersect.

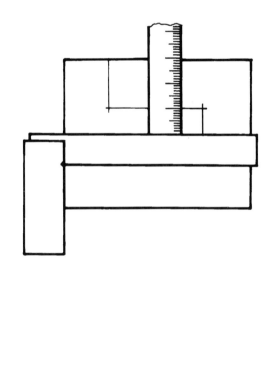

Example 2: component shown in Fig. 5.27

The plate shown at step 1 has been cut out 2 mm oversize on length and width and has not been filed. All four sides have sawn edges.

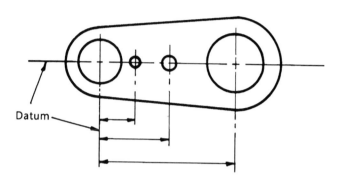

Figure 5.27 Component used in Example 2.

Step 1 Measure from each long edge and find the centre using a precision steel rule. Scribe the centre line using the edge of the rule as a guide. Find the centre of the small radius by measuring from one end the size of the radius plus 1 mm (this allows for the extra left on the end). Centre dot where the lines intersect.

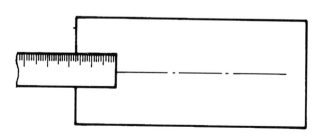

Step 2 Using dividers, set the distance from the centre of the small radius to the centre of the first small hole. Scribe an arc. Repeat for the second small hole and the large radius. Centre dot at the intersection of the centre lines. The dividers are set using the graduations of a precision steel rule.

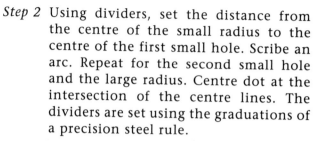

Step 3 Set dividers to the small radius. Locate on the centre dot and scribe the radius. Repeat for the large radius and if necessary, the two holes.

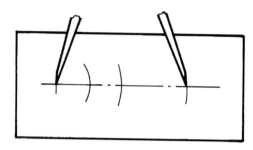

Step 4 Complete the profile by scribing a line tangential to the two radii using the edge of a precision steel rule as a guide.

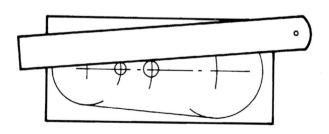

Example 3: component shown in Fig. 5.28

The plate shown at step 1 has been roughly cut to size and requires complete marking out of the profile and holes.

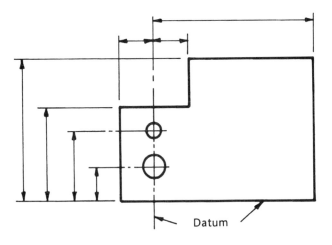

Figure 5.28 Component used in Example 3.

Step 1 Clamp the plate to the face of an angle plate, ensuring that the clamps will not interfere with marking out. Use a scriber in a surface gauge and set the heights in conjunction with a precision steel rule. Scribe the datum line. Scribe each horizontal line the correct distance from the datum.

Step 2 Without unclamping the plate, swing the angle plate on to its side (note the importance of clamp positions at step 1). This ensures that the lines about to be scribed are at right angles to those scribed in step 1, owing to the accuracy of the angle plate. Scribe the datum centre line. Scribe each horizontal line the correct distance from the datum to intersect the vertical lines.

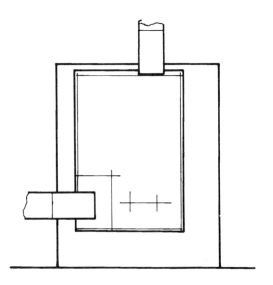

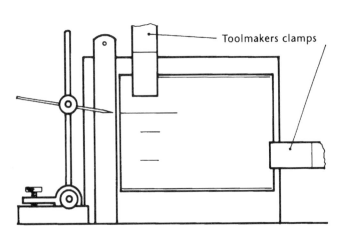

Example 4: component shown in Fig. 5.29

The plate shown at step 1 is to be produced from the correct width (*W*) bright rolled strip and has been sawn 2 mm oversize on length (*L*). The base edge has been filed square to the ends and it is required to mark out the angled faces.

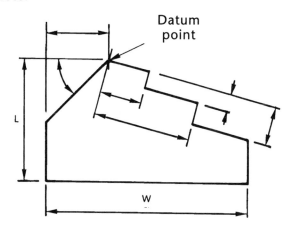

Figure 5.29 Component used in Example 4.

Step 1 Using a precision steel rule measure from two adjacent edges to determine the datum point. Centre dot the datum point.

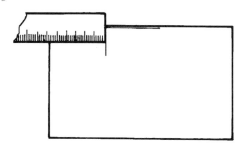

Step 2 Set protractor at required angle and scribe line through datum point.

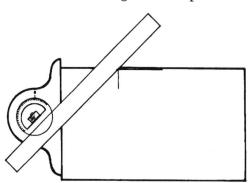

Step 3 Reset protractor at second angle and scribe one line through datum point Scribe the remaining two lines parallel to and the correct distance from the first line using the protractor at the same setting.

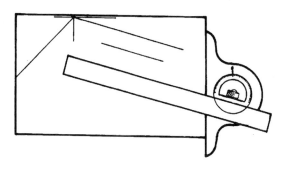

Step 4 Set dividers at correct distances, locate in datum centre dot and mark positions along scribed line.

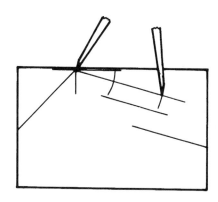

Step 5 Reset protractor and scribe lines through marked positions.

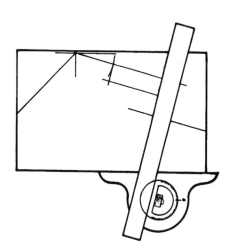

Example 5: shaft shown in Fig. 5.30

The shaft shown is to have a keyway cut along its centre line for a required length. Accurate machining is made possible by setting up the shaft relative to the marked out position.

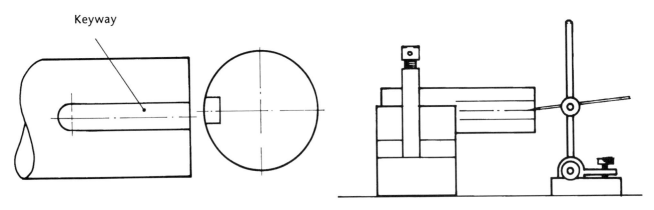

Figure 5.30 Component used in Example 5.

Step 2 Clamp shaft in a vee block ensuring that the line marked at step 1 is lying horizontal. This can be checked using a scriber in a surface gauge. Transfer the centre line along the required length of shaft. Scribe two further lines to indicate the width of slot.

Step 1 Scribe line on the end face through the centre of the shaft using the centre head of a combination set (see Fig. 5.17).

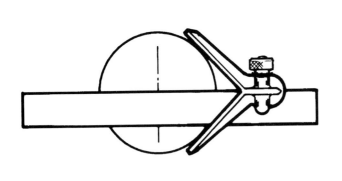

Step 3 The required length of slot can be marked without removing it simply by turning the vee block on its end and scribing a horizontal line at the correct distance from the end of the shaft.

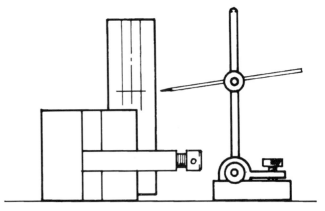

6

Work and tool holding

Degrees of freedom

Any body in space possesses six degrees of freedom as shown in Fig. 6.1. The body can have linear motion along any one of three mutually perpendicular axes X, Y, and Z, or it may have rotational motion about any one of these three axes. The relationship of these three axes to a machine tool is shown in Fig. 6.2. The branch of science relating to motion is known as kinematics and the aim of kinematic design is to prevent movement in certain degrees of freedom by applying constraints, i.e. by controlling or restricting motion by applying a force.

For example, the design of a machine tool is such that each element is restricted from moving in any direction other than the one required. To be effective a machine tool must provide for linear motion by means of slides and for rotational motion by means of spindles all accurately aligned relative to each other.

It follows that if a workpiece is to have machining operations carried out upon it then its movement too must be controlled and in the same way cutting tools mounted in the machine tool must be restricted.

Workholding

If work is to be machined, then all degrees of freedom must be controlled so that there is no possibility of the workpiece moving due to the forces exerted during the machining operation.

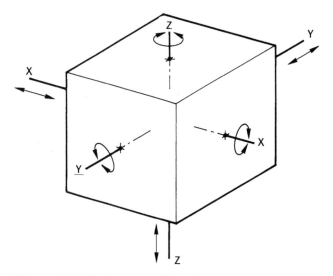

Figure 6.1 Six degrees of freedom.

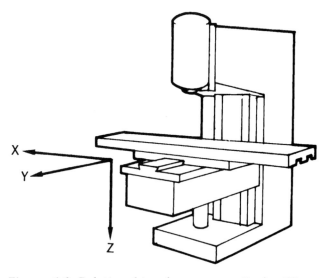

Figure 6.2 Relationship of axes on vertical milling machine.

The workpiece may need to be located in the correct position relative to the machine and its cutting tool. For a few items, simple location methods may be used. For batch production, jigs or fixtures may be required, but this is beyond the scope of this book.

It should be noted here that work should never be held by hand while it is being machined – **always** use the appropriate work holding device; **never** take a chance.

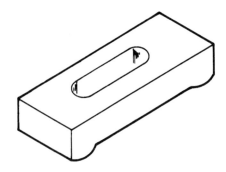

Figure 6.4 Clamp.

Clamping

One of the simplest forms of holding work is to clamp the workpiece directly to the machine table using machine clamps. The same principles apply whether you are using a drilling or milling machine. Remember, the purpose is to prevent the work from being able to move during machining. In most cases this will be achieved using two clamps. To operate safely, two clamps is the minimum you should use.

Tee-slots are provided in machine worktable surfaces into which are fitted tee-bolts or tee-nuts in which studs are screwed (see Fig. 6.3). Tee bolts and studs are available in a variety of lengths to suit a wide range of workpiece thicknesses. Various styles and shapes of machine clamp are available, one of which is shown in Fig. 6.4. The central slot allows it to be adjusted to suit the workpiece. To provide sound clamping, the clamp should be reasonably level and this is achieved by packing under the rear of the clamp as near as possible the same height as the workpiece (Fig. 6.5). The tee bolt should be placed close to the work, since the forces on the work and the packing are inversely proportional to their distances from the bolt, i.e. for greatest clamping force on the work, distance *A* in Fig. 6.5 must be less than distance *B*.

When clamping a workpiece directly on the machine worktable, care must be taken to avoid damage to the worktable, e.g. when drilling holes through the workpiece and on through the worktable. To avoid this it may be necessary to raise the work off the worktable on parallels and clamp over the parallels as shown in Fig. 6.6.

Care must be taken when clamping as considerable forces are exerted which may cause damage to the workpiece. Always clamp at thick sections of the workpiece never on thin or unsupported sections. If the workpiece is thin or unsupported, then suitable support or packing must be used (Fig. 6.7).

Vices

Small parts can be conveniently and securely held in a vice similar to that shown in Fig. 6.8. The vice should always be securely clamped to the machine worktable.

Repeat positioning of work in the vice can be achieved using the end-stop on the vice shown in Fig. 6.8.

Vices can be conveniently used on drilling and milling machines and again care should be taken to avoid machining the vice at the same time as the workpiece.

Take care when tightening the vice as too much force can distort or damage the workpiece or lead to damage of the vice itself.

If work needs to be accurately located on a milling machine, the vice needs to be lined up relative to one of the machine slide movements. This is done by attaching a dial

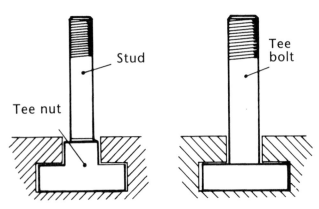

Figure 6.3 Tee nut and tee bolt.

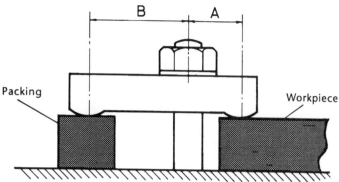

Figure 6.5 Clamping forces.

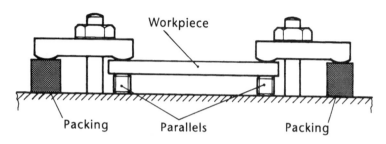

Figure 6.6 Workpiece clamped on parallels for support.

indicator to a fixed part of the machine and checking along the length of a parallel held in the vice. For example, if the jaws require to be parallel to the table movement, the dial indicator is fixed to the column of the machine and the table, vice and the parallel held in it, is moved, by hand, past the dial indicator (Fig. 6.9). The vice is adjusted until a constant reading is obtained in the dial indicator, and the vice is then securely clamped down. A parallel is used for this operation in order to give the same condition as a clamped workpiece. The surface of the parallel is straight and undamaged, and lengths longer than the vice jaws can be used to give greater accuracy of setting.

Vee blocks

These are used either singly or in pairs, for cylindrical work. Where a hole is to be drilled at right angles to the axis of a cylindrical piece, the position of the hole would be marked out (Chapter 5) and the work set up relative to the marking out and clamped in position before drilling (Fig. 6.10).

Vee blocks can also be used in conjunction with a vice for milling operations. The vee block and the cylindrical workpiece are set vertically in the vice (Fig. 6.11) which is then tightened. The required machining operation can then be carried out on the workpiece. Take

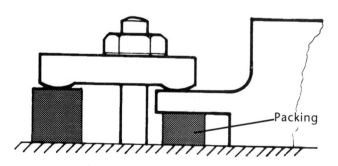

Figure 6.7 Packing under unsupported work.

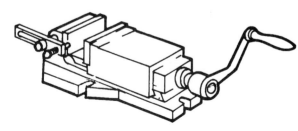

Figure 6.8 Machine vice.

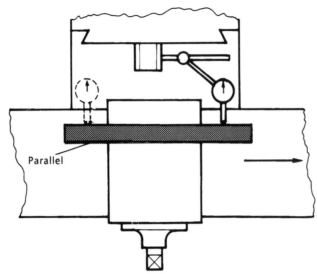

Figure 6.9 Lining up a vice.

care when tightening the vice especially with cast iron vee blocks, as it is possible to split the vee block if too much force is applied.

Self-centring chuck

Self-centring chucks are used on centre lathes and restrict the work in all but rotation about the spindle axis. The most common is the three-jaw self-centring type (see Fig. 6.12) which is used to hold circular or hexagonal workpieces.

The chuck operates by means of a pinion engaging in a gear on the front of which is a scroll, all encased in the chuck body. The chuck jaws, which are numbered and must be inserted in the correct order, have teeth which engage in the scroll and are guided in a slot in the face of the chuck body. As the pinion is rotated by a chuck key, the scroll rotates, causing all three jaws to move simultaneously and so automatically centre the work.

Four-jaw self-centring chucks for use with square bar or square workpieces are also available.

Between centres

A workpiece having a number of diameters which are required to be concentric, can be machined between centres.

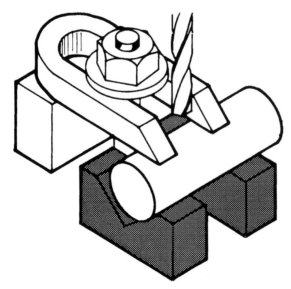

Figure 6.10 Vee blocks used in drilling.

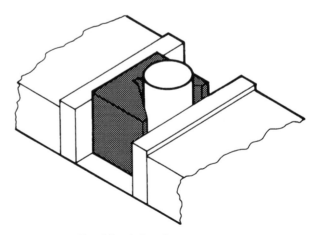

Figure 6.11 Vee block in vice.

Figure 6.12 Three-jaw self-centring chuck.

The workpiece must have locations (or centres) at each end in order to locate with the machine centres. These locations are put into the end of the workpiece using a centre drill (Fig. 6.13a) and the machine centres then locate in the seating as shown in Fig. 6.13(b).

A centre is inserted in the spindle nose which rotates with the spindle and workpiece, and is referred to as a 'live centre'. A centre

inserted in the tailstock is fixed, does not rotate with the workpiece and is referred to as a 'dead' centre. Great care must be taken to prevent overheating of 'dead' centres due to lack of lubrication or too high a pressure. Keep the centre well lubricated with grease, and do not overtighten the tailstock.

Revolving centres, where the centre runs in bearings and rotates with the workpiece are available and remove the problems likely to be encountered with 'dead' centres.

In order to drive the workpiece, a work driver plate must be mounted on the spindle nose and the drive completed by attaching a work carrier to the workpiece, Fig. 6.14. This method locates the work on the centres and again restricts the work in all but rotation about the spindle axis.

Faceplate

A faceplate is used on a centre lathe for workpieces which cannot be easily held by any of the other methods. When fixed to the spindle nose of the lathe, the face of the faceplate is square to the axis of spindle rotation. A number of slots are provided in the face for clamping purposes. Workpieces can be clamped directly to the faceplate surface but, where there is a risk of machining the faceplate, the workpiece must be raised from the surface on parallels before clamping as previously explained. Positioning of the workpiece depends upon its shape and the accuracy required.

Flat plates which require a number of holes are easily positioned by marking out the hole positions and using a centre drill in a drilling machine to centre each position. A fixed centre in the tailstock is then used to locate the centre position and hold the workpiece against the faceplate while clamping is carried out (Fig. 6.15).

Workpieces which already contain a hole which is to be enlarged, e.g. cored holes in a casting, can be marked out to produce a box in the correct position, the sides of which are the same length as the diameter of the required hole. Roughly positioned and lightly clamped, the workpiece can be set accurately using a scriber in a surface gauge resting on the cross

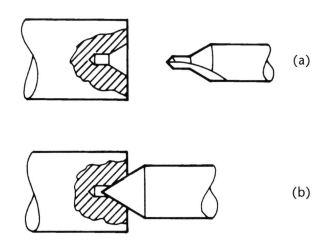

(a)

(b)

Figure 6.13 Centre drill and seating for machine centres.

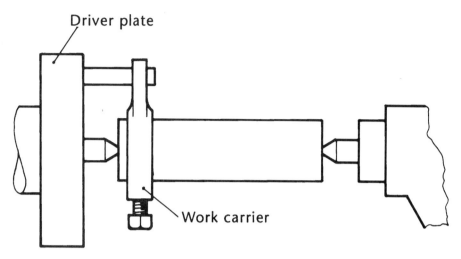

Figure 6.14 Workpiece between centres.

slide surface. The faceplate is rotated by hand and the workpiece is tapped until all four scribed lines are the same height indicating that the hole is on centre (Fig. 6.16). The workpiece is then securely clamped.

When the workpiece has been clamped, check each nut and screw to ensure it is tight. Turn the faceplate by hand and check that all bolts, clamps and work is clear of the bed, cross slide, toolpost and tool. To help with this, avoid using excessively long clamping bolts. Check for 'out of balance' of the faceplate – a counterbalance may be required.

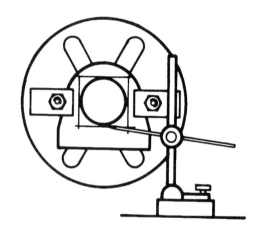

Figure 6.16 Setting workpiece on faceplate.

Angle plate

Where a machining operation has to be carried out parallel to a face, the workpiece movement can be controlled using an angle plate. The workpiece is clamped to the angle plate and when located in the correct position, the angle plate is clamped to the machine table (Fig. 6.17). The operation of milling or drilling can then be carried out.

Small angle plates can be used in conjunction with a face plate on the centre lathe, where the datum and machining operation are parallel to the axis of rotation (Fig. 6.18). The angle plate is first roughly positioned on the face plate and the workpiece clamped to the angle plate. The set-up is then adjusted to bring the workpiece into the required position

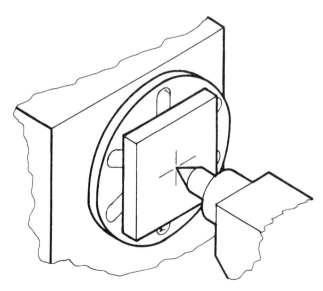

Figure 6.15 Locating workpiece on faceplate.

the same way, cutting tools must be located and constrained so that they too can carry out the machining operation safely and accurately.

Drills

Twist drills are available with parallel shanks up to 30 mm diameter and with taper shanks up to 100 mm diameter and are made from high-speed steel.

Twist drills and similar tools with parallel shanks are held in a drill chuck (Fig. 6.19). Many different types of chuck are available, each being adjustable over its complete range, and give good gripping power. The twist drill is thus located and restricted in all but rotation about the spindle axis.

By rotating the outer sleeve, the jaws which are self-centring, can be opened and closed. To ensure maximum grip, the chuck should be tightened using the correct size of chuck key. This prevents the drill from spinning in the chuck during use and chewing up the drill shank.

The chuck is fitted with a Morse-taper shank which fits into a corresponding Morse taper in the spindle again providing the necessary location and control. The size of Morse taper is identified from smallest to largest by the num-

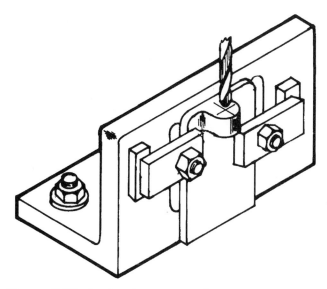

Figure 6.17 Angle plate on machine table.

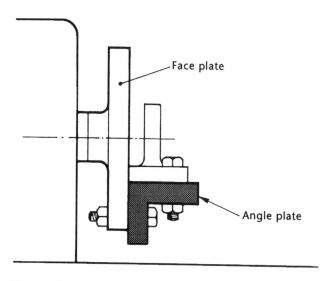

Figure 6.18 Angle plate on faceplate.

for machining and both angle plate and workpiece securely clamped. Great care is needed with this type of set up since the complete assembly may be out of balance in which case counterbalance weights will need to be added to the face plate.

Toolholding

We have seen how workpieces are located and controlled in order to have machining operations carried out as required and in safety. In

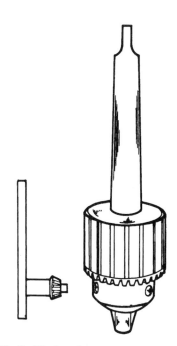

Figure 6.19 Drill chuck.

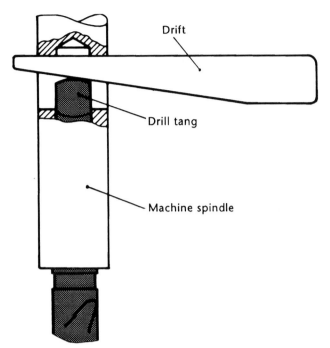

Figure 6.20 Drift in drill spindle.

Figure 6.21 Taper shank drill.

bers 1, 2, 3, 4, 5, and 6. The included angle of each taper is different but is very small, being in the region of 3°. If the two mating tapered surfaces are clean and in good condition, this shallow taper is sufficient to provide a drive between the two surfaces. At the end of the taper shank, two flats are machined, leaving a portion known as the tang. This tang fits in a slot on the inside of the spindle and its purpose is for the removal of the shank.

To remove a shank from the spindle, a taper key known as a drift is used. The drift is inserted through a slot in the machine spindle as shown in Fig. 6.20 and levered or the end tapped with a hammer to release the shank.

Drills are available with Morse-taper shanks which fit directly into the spindle without the need for a chuck (Fig. 6.21). The size of Morse taper depends on the drill diameter and the range is shown in Table 6.1.

It is essential that tapers are kept clean and in good condition. As already stated, the drive is by friction through the tapered surfaces, and any damage to these surfaces put some of the driving force on the tang. If this force is excessive, the tang can be twisted off. When this happens the drill has to be discarded, as there is no way of easily removing it from the spindle.

Where a cutting tool or chuck has a Morse taper smaller than that of the machine spindle, the difference is made up by using a sleeve. For example, a drill with a No. 1 Morse-taper shank to be fitted in a spindle having a No. 2 Morse taper would require a 1–2 sleeve, i.e. No. 1 Morse-taper bore and a No 2 Morse taper outside. Sleeves are available from 1–2, 1–3, 2–3, 2–4, and so on over the complete range.

Cylindrical milling cutters

A cylindrical cutter, sometimes referred to as a plain milling cutter, has teeth on the periphery only, and is used to produce flat surfaces parallel to the axis of the cutter (Fig. 6.22a). The teeth are helical, enabling each tooth to take a cut gradually, thereby reducing shock and minimising chatter. Cylindrical cutters are made in a variety of diameters and lengths up to 100 mm diameter × 150 mm long.

The cutter has a hole through the centre for mounting on an arbor. A standard arbor for

Table 6.1

Morse taper		No. 1	No. 2	No. 3	No. 4	No. 5	No. 6
Drill diameter range (mm)	above		14	23	32	50	76
	up to	14	23	32	50	76	100

use in a horizontal milling machine is shown in Fig. 6.23. One end has an international taper to suit the machine spindle, for location. A threaded hole in the end provides the means of holding the arbor in position, by means of a drawbolt through the machine spindle. The flange contains two key slots to provide the drive from two keys on the spindle nose. The long diameter is a standard size, to suit the hole size of the cutter, and the thread carries the arbor nut to clamp the cutter. A keyway is cut along the length of this diameter into which a key is fitted, to provide a drive and prevent the cutter slipping when taking heavy cuts. To position the cutter along the length of the arbor, spacing collars are used. These are available in a variety of thicknesses, with the ends ground flat and parallel. Towards the end, a larger bush is positioned. This has an outside diameter to run in the bearing of the arbor support and is known as the 'running bush'.

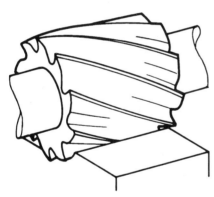

Figure 6.22(a) Cylindrical cutter.

End milling cutter

An end milling cutter (see Fig. 6.22b) has a number of helical teeth on its circumference and teeth on one end and is used for light operations such as milling slots, profiling, and facing narrow surfaces. The end teeth are not cut to the centre, so this type of cutter cannot be fed in a direction along its own axis.

A similar cutter known as a slot drill (Fig. 6.22c) has two helical teeth on its circumference and two teeth on the end, cut to the centre and can be fed along its own axis in the same way as a drill. The slot drill is used to produce keyways and blind slots with the cutter sunk into the material like a drill and then fed longitudinally the length of the keyway or slot, cutting on its circumference.

Both types of cutter have a screwed shank which is mounted in a special chuck (Fig. 6.24) to provide the necessary location and constraint. The chuck body has an appropriate taper shank to suit the machine, usually an international taper, and has an internal collet and locking sleeve. The collet, which is split along the length of its front end, is internally threaded at its rear end to suit the cutter. Collets with different bore sizes are available to suit the shank diameter of the cutter being used. The collet is inserted into the locking sleeve and the assembly screwed into the chuck body until the flange almost meets the end face of the body.

The cutter is inserted and screwed into the collet until it locates on a centre inside the chuck body and becomes tight. The centre

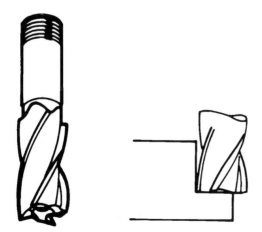

Figure 6.22(b) End milling cutter.

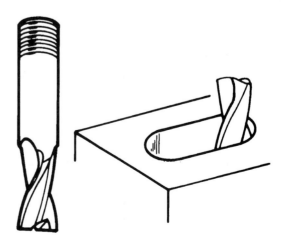

Figure 6.22(c) Slot drill.

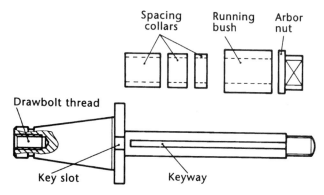

Figure 6.23 Standard milling machine arbor.

locates the end of the cutter and ensures rigidity and true running. A spanner is used on the locking sleeve to give the collet a final tighten.

The cutter is now located and held by the collet and the centre and totally restricted. The cutter cannot push in or pull out during the cutting operation. Any tendency of the cutter to turn during cutting will tighten the collet still further and increases its grip on the cutter shank. This type of collet chuck is located, restricted and driven in the machine spindle in the same way as the previously mentioned arbor.

Single point cutting tool

Single point lathe tools are located and restricted in a toolpost mounted on the top slide of the centre lathe. The lathe tool has to be set exactly on centre of the machine spindle and this is done by inserting the correct thickness of packing under the tool. Two or three screws are tightened on top of the tool to secure it in position (Fig. 6.25). The lathe tool is now fixed to, and becomes part of the slide assembly, which, due to the construction of the centre lathe is restricted to move parallel and at right angles to, the spindle axis. The tool, now being part of this assembly is thus controlled to move in the same way. Thus, if the tool is made to move parallel to the axis of a rotating workpiece while removing material, it will produce a parallel circular cylinder (Fig. 6.26a). Likewise, if the tool is moved at right angles to the axis of rotation of a rotating workpiece while removing material, it will produce a flat surface (Fig. 6.26b).

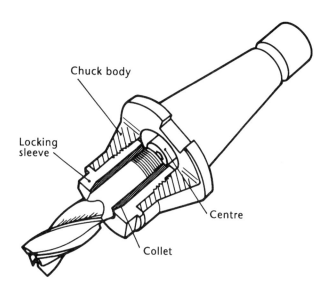

Figure 6.24 Milling chuck for screwed shank cutters.

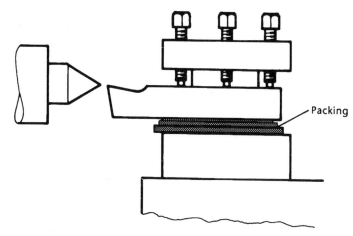

Figure 6.25 Lathe tool with packing in position.

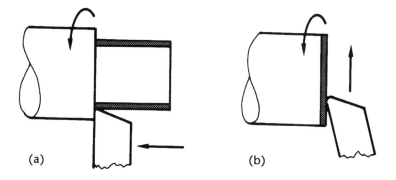

Figure 6.26 Surfaces produced by lathe tool.

Index